AIRFLOW IN
DUCTS

by

LEO A. MEYER

H. LYNN WRAY, P.E., Technical Advisor

**INDOOR ENVIRONMENT
TECHNICIANS LIBRARY**

FOREWORD

You are probably working as a technician in one of the indoor environment fields. This means that you have at least some understanding of airflow in ducts. However, don't fall into the trap of thinking, "I know all this stuff."

Read each chapter. Then do the Review. In my experience, every time I studied material I "knew all about," I learned new ideas and corrected misunderstandings.

If you study each chapter carefully, you will gain new ideas. More important, you will give yourself a solid understanding of basic principles that you will be able to apply in the field. You will also be able to apply your knowledge to more advanced technical principles covered in later books in this series.

Indoor Environment Technician's Library

This book is part of the *Indoor Environment Technician's Library*. These are practical books that you can use for training or as reference. These books apply to all areas of the indoor environment industry:

- Heating, ventilating, and air conditioning
- Energy management
- Indoor air quality
- Service work
- Testing, adjusting, and balancing

If You are Training Others

If you are a supervisor training others, you will find that the *Indoor Environment Technician's Library* can make it easier. A Supervisor's Guide is available for each book. It includes teaching suggestions and key questions you can ask to make sure the student understands the material.

Leo A. Meyer

LAMA Books
Leo A. Meyer Associates Inc.
23850 Clawiter Road
Hayward CA 94545-1723
510-785-1091

ISBN 0-88069-018-6

© 1996 by Leo A. Meyer Associates Inc.
All rights reserved. No part of this publication may be reproduced, stored in an electronic retrieval system or transmitted in any form or by any means—electronic, mechanical, or otherwise—without the prior permission of the publisher.

Leo A. Meyer Associates, Inc. specifically disclaims any and all liability for damages of any type whatsoever that may result directly or indirectly from a person's reliance upon or utilization of the information contained in this book.

TABLE OF CONTENTS

1. Basics of Airflow — 1
2. Calculating Duct Sizes — 11
3. Air Quantity and Velocity — 26
4. Pressures in a Duct — 38
5. Airflow in a Duct and Dynamic Losses — 49
6. Sizing Ductwork — 58
7. Calculating Pressure Losses in Ductwork — 71
8. Duct Fittings — 83
9. Measuring Airflow — 97

Review Answers — 108

Appendix—Equations Used in This Book — 116

Index — 119

1  BASICS OF AIRFLOW

If you work in the HVAC industry, you are concerned with the process of conditioning air and moving air from one area to another. It makes sense that you need to know how air flows in ducts and what factors affect the flow.

When you complete this book, you will understand airflow in a duct better than most people in the HVAC industry do. This will help you fabricate and install ductwork more effectively and help you make changes in ductwork size with confidence. A good understanding of airflow is needed for different specialties in the HVAC industry, such as:

- ☐ Service work
- ☐ Testing, adjusting, and balancing
- ☐ Energy management
- ☐ Indoor air quality work

WHY MOVE AIR?

Think of what HVAC means—heating, ventilating, and air conditioning:

- ☐ **Ventilating** is bringing outside air into a building.
- ☐ **Heating and air conditioning** means heating, cooling, and cleaning air, and regulating its moisture content.

Conditioned air must be delivered to selected areas of a building and then removed from those areas and returned for re-conditioning. Conditioned air is usually transported through ductwork.

The process of moving air in ducts distributes energy:

- In cold weather, heat taken from an energy source such as gas or electricity is added to air. This heat is delivered to the conditioned space.

- In hot weather, heat is removed from air by the use of electrical energy. In this case, heat energy is being removed from the conditioned space.

THE HVAC SYSTEM

An HVAC system has different components:

- **Central air handling system**—Generally contained in a mechanical room. It includes the fan to move the air and equipment to condition the air before it is delivered to the conditioned space.

- **Boiler**—Provides hot water to heat the air.

- **Refrigeration unit**—Provides a means to cool the air.

- **Duct system**—Distributes the conditioned air where needed.

Central Air Handling System

Figure 1 shows a typical central air handling system. This is a schematic drawing that shows the parts of the system and how they are related to each other. It does not show the location of the components in an actual installation. There are many different system variations. You need to understand the basic components of a system and their relationship to each other. Then you can identify the components on any job.

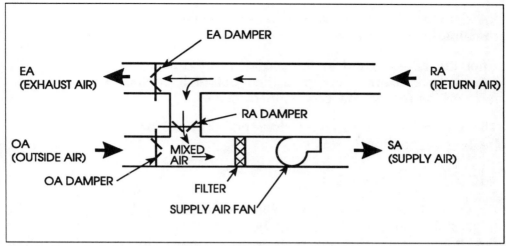

Fig. 1: Central air handling system

Supply air (SA) is the conditioned air delivered to the building. In Fig. 1 the supply air is in the lower right hand corner.

When supply air is delivered into a room, an equal amount of air must be removed from the room. This **return air (RA)** is transported back to the central air system for reprocessing. In Fig. 1 the return air is in the upper right corner.

Only a percentage of the return air can be reused. The air would become stale if the same air were used over and over again. To avoid this, fresh air is brought into the system through the **outside air (OA)** intake. In Fig. 1 the outside air inlet is in the lower left.

When outside air enters the building, the same amount of air must be removed from the building through the **exhaust air (EA)** outlet or other exhaust air systems.

The exhaust air (EA), outside air (OA), and return air (RA) ducts all operate together. An automatic control system operates the damper motors to maintain the proper mix of air. The OA and the EA dampers open together and close together to balance the air entering and leaving the building. As the OA damper opens, the

RA damper closes so that the same amount of air remains in the system.

After the air is mixed in the proper proportions, the **mixed air** is drawn through a filter before it enters the fan and returns to the conditioned spaces.

The system shown in Fig. 1 does not provide for heating or cooling the air.

Heating System

Figure 2 shows how heat is added. A **heating coil** (Fig. 3) is added to the system before the air enters the fan. It is located after the filter so that there is less chance of dirt clogging the coil.

The coil is similar to a car radiator. Hot water flows through tubes in the coil. Metal fins are attached to the

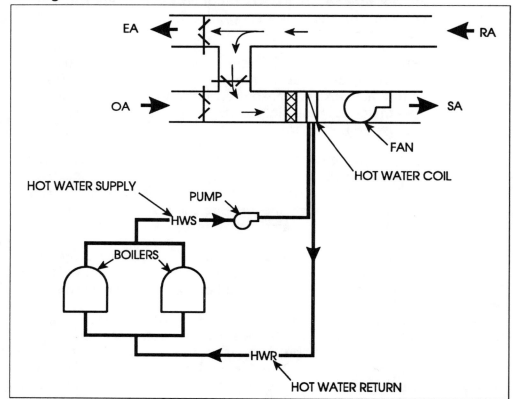

Fig. 2: Hot water system added to air handling system

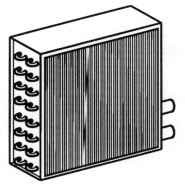

Fig. 3: A heating coil

tubes. The heat from the hot water is conducted to the outside of the tubes and to the fins. The air passing across the coil is heated by absorbing heat from the fins and the tubes.

The hot water for the coil is heated by a **boiler**. A pump circulates the hot water through the coil.

Cooling System

Figure 4 shows a **chilled water coil** added to the system to cool the supply air. The chilled water coil is similar to the hot water coil. Chilled water flows through the tubes to cool the air.

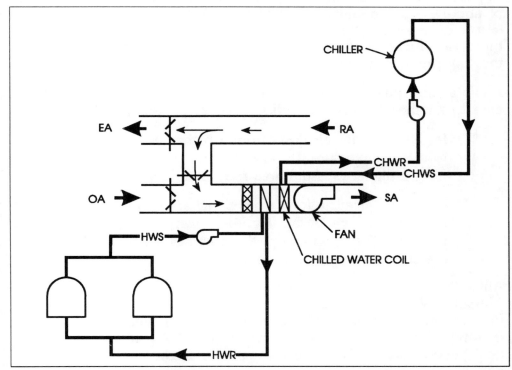

Fig. 4: Chilled water system added

The water is cooled by a **chiller,** which is a refrigeration unit. A pump circulates water through the chiller where it is cooled and then returned to the chilled water coil.

The system shown in Fig. 4 is a complete air handling system. It can:

- Mix the return air and outside air to the correct proportions.
- Filter the air.
- Heat the air.
- Cool the air.

WHY AIR FLOWS

A water system in a house is usually under a pressure of about 30 **psi (pounds per square inch).** When a faucet on the water line is opened, the line pressure at the open faucet decreases to zero. The water flows to the low pressure area.

Air is a **fluid** just like water. It also flows from one area to another because of a **difference in pressure:**

- In the open air, air flows from a higher pressure to a lower pressure. Wind is air that is moving from a higher pressure area to a lower pressure.
- In a duct, air also flows from a higher pressure to a lower pressure. A fan (Fig. 5) creates the higher pressure. The open end of the duct has a lower pressure, so the air flows out.

All air in an open system is under **atmospheric pressure,** which is normally 14.7 psi at sea level. Pressure in a duct refers to the pressure that is higher or lower than the atmospheric pressure of 14.7 psi. A positive pressure (+) is above atmospheric pressure. A negative pressure (−) is below.

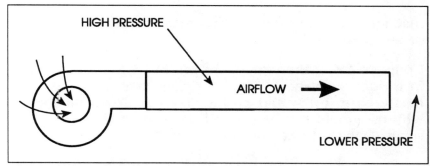

Fig. 5: The fan creates pressure that causes air to flow

The amount of air flowing through a duct is regulated by the amount of **pressure difference** and by the **system resistance**. The higher the pressure difference, the greater the air velocity and the greater the quantity of air that will flow from the duct.

Friction is a resistance which slows down airflow. The flow of air creates friction as it rubs against the side of the duct, and the friction creates resistance to the airflow. Think of blowing through a piece of garden hose 6 inches long. You can feel a good stream of air coming out of the tube. Now try to blow through a hose 50 feet long. Little or no air comes out the other end. This is because friction created by the sides of the long hose reduces the pressure at the open end of the hose.

MEASURING DUCT PRESSURES

Since the air pressure in a duct directly affects the flow of air, measuring the air pressure is important. Air pressure in a tire is measured in **psi (pounds per square inch)**. Air pressure in a duct is measured in **inches water gage (inches wg or "wg)**. Both psi and inches wg measure the same thing—the amount of pressure on a given amount of area. However, inches wg is used for duct because it is suitable for measuring small values. Compare the measurements taken by a carpenter and a machinist. The carpenter uses a rule divided into eighths and sixteenths of an inch. The machinist uses instruments that measure in thousandths of an inch.

Compare the measurements of psi and wg. The pressure in a duct might be 1" wg. This same duct pressure measured in psi would be just 0.04 psi. You can see why psi would not be a good scale for measuring small changes in air pressure in ducts.

THE WATER GAGE

The term **inches wg** (or **"wg**) means **inches of water differential in a water gage.** A **water gage** (Fig. 6) is the basic device for measuring air pressure in duct. In actual practice, other instruments that are more convenient to use, such as an **electronic manometer** (Fig. 7), are used to measure "wg. These will be described in another book in this series. But all instruments measuring inches wg are based on the principle of the water gage, which is explained in Figs. 8 and 9.

Figure 8 shows a small, flexible, plastic hose connected to a U-shaped glass tube at B. The pressure at the open ends of the tubes (A and E) is equal (because it is only atmospheric pressure). Therefore the level of the water at C and D is the same.

Figure 9 shows the open end of the hose placed in a duct in which the air is flowing. Therefore the pressure at A is greater than the pressure at E. This pressure pushes the water level down at C. This means that the water level at D must rise by the same amount.

Fig. 6: A water gage

Fig. 7: An electronic manometer

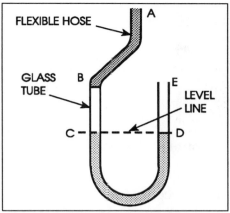

Fig. 8: U tube with equal pressure on both ends

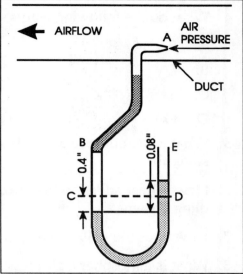

Fig. 9: U tube with "wg pressure reading

If the pressure pushes the water down 0.4" at point C, the level must rise 0.4" at point D. Therefore there is a difference of 0.8" between the two levels (Fig. 9). The pressure reading on this U-tube is 0.8" wg.

REVIEW

If you can answer the following questions without referring to the text, you have learned the contents of this chapter. Try to answer every question before you check the answers in the back of the book.

1. What do the letters HVAC stand for?

2. List the four basic components of an HVAC system.

3. How is the conditioned air usually transported from the central HVAC to the conditioned space?

4. What do the following letters stand for?
   A. RA
   B. EA
   C. SA
   D. OA
5. What is the purpose of a coil in a central air handling system?
6. How is heat added to the mixed air in a central air handling system?
7. How does the chilled water coil cool the air?
8. What is atmospheric air pressure at sea level in psi?
9. Why is duct pressure measured in " wg instead of psi?
10. What does wg stand for?
11. What would a pressure of 1" wg do to the water level of a U tube?

2 CALCULATING DUCT SIZES

USING MATH

To work with airflow in ducts you need to know how to work simple equations and how to change equations to a different form. These two skills are essential for all HVAC technicians. If you need to review this basic math, use another book in this series entitled *Math for the Indoor Environment Technician*. This book is direct and easy to understand. It reviews only the kind of math needed on the job. It will give you skills that will make the rest of your studies much easier.

This chapter assumes that you know how to use simple equations and know how to use a calculator. You will use equations to calculate duct sizes (covered in this chapter) and air quantity and velocity (covered in the next chapter). These and many other calculations are important for any HVAC technician.

You should have a calculator that has a key for pi (π) and a key for square root:

$\boxed{\pi}$ $\boxed{\sqrt{}}$

It is also useful to have keys for parentheses, marked:

$\boxed{(}$ $\boxed{)}$

For some industry areas, such as service work, energy management, and TAB (testing, adjusting, and balancing), it is also useful to have a calculator with a key that gives cube roots:

Cube roots are not needed for this book.

WORKING WITH DUCT SIZES

Calculating duct sizes is very important because the quantity and velocity of air delivered by a system are related to the duct size. You may need to calculate what size duct is needed to deliver a given amount of air. Or you may have to calculate what size rectangular duct has the equivalent area of a given round duct. To work with duct sizes, you must be able to do the following:

- ❑ Calculate the area of a rectangular duct when you know the duct dimensions.
- ❑ Convert square inches to square feet and convert square feet back to inches.
- ❑ Find the dimension of one side of a rectangular duct when the duct area and one side are known.
- ❑ Calculate the area of round duct when the diameter is known.
- ❑ Calculate the diameter of round duct when the area is known.

All of these operations are explained in this chapter. Each one is a simple operation, especially if you use a calculator. Be sure that you understand each of them before leaving the chapter. You will need all of these in order to work the problems with air quantity and velocity in the next chapter and in your work.

FIND THE AREA OF DUCT

The cross section of rectangular duct is a rectangle (Fig. 1). You often have to find the **area** of the

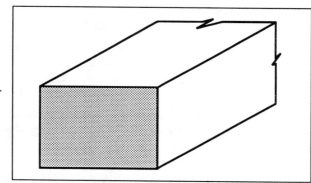

Fig. 1: Cross section of a duct

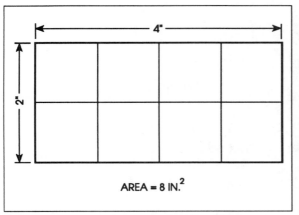

Fig. 2: The area of a rectangle 2" x 4" is 8 square inches

cross section. To find the area of any rectangle, multiply the dimensions of one side by the other. For example, to find the area of the rectangle in Fig. 2, multiply the height (2 inches) by the width (4 inches):

Area = Width x Height

Area = 4" x 2"

Area = 8 square inches

As you can see in Fig. 2, there really are 8 boxes one inch square in the rectangle that is 4" x 2".

Use this procedure to find the area of a duct. Since duct sizes are given in **inches,** if you multiply one side by the other side you have the area in **square inches.** The abbreviation is **sq. in.** or **in^2.**

NOTE: When identifying rectangular duct, the width (horizontal) is given first. The height (vertical) is given second. For duct drawn in an HVAC system, the dimension of the side shown in the drawing is given first. Thus a duct could be designated as 18" x 12" or 12" x 18" depending on the view of the drawing. Dimensions for rectangular duct are usually even numbers. **Even numbers** can be divided by 2 (for example, 8, 10, 12, 14, etc.).

EXAMPLE

Find the area of a duct that is 10" x 8":

Area = Width x Height

Area = 10" x 8"

Area = 80 sq. in.

PROBLEMS

Do these problems before you go on. After you work the problems, check the answers at the end of this chapter.

1. A duct is 30" x 24". What is its area in square inches?

2. What is the area of an 18" x 18" duct in square inches?

CHANGE SQUARE INCHES TO SQUARE FEET

For most problems in HVAC, you need to know duct areas in **square feet** rather than **square inches**. Air velocity is measured in **feet per minute** and air quantity is measured in **cubic feet per minute**. Therefore, you need to know duct area in square feet in order to use it with these other values.

A square foot measures one foot on each side of a square. That means it is 12 inches on each side (Fig. 3). Therefore the area of one square foot is 144 square inches:

Area = Width x Height

Area = 12" x 12"

Area = 144 sq. in.

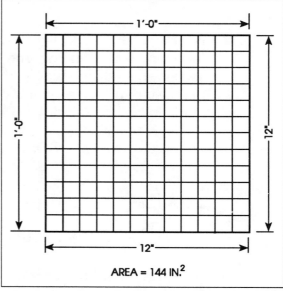

Fig. 3: One square foot equals 144 square inches

Since 144 sq. in. is the same as one square foot, to change **square inches** to **square feet** (sq. ft. or ft^2), divide by 144:

$$\text{sq. ft} = \frac{\text{sq. in.}}{144}$$

EXAMPLE

A rectangular duct measures 24" x 18". What is its cross-sectional area in square feet?

Step 1: Find the area in square inches.

Area = Width x Height

Area = 24" x 18"

Area = 432 sq. in.

Step 2: Find the area in square feet.

$$\text{Area in sq. ft.} = \frac{\text{sq. in.}}{144}$$

$$\text{Area in sq. ft.} = \frac{432 \text{ sq. in.}}{144}$$

Area = 3 sq. ft.

These two math steps can be combined into one equation:

$$\text{Area} = \frac{\text{Width x Height}}{144}$$

$$\text{Area} = \frac{24" \text{ x } 18"}{144"}$$

Area = 3 sq. ft.

NOTE: This problem can also be worked by changing 24" and 18" to feet before multiplying them. Do this by dividing the inches by 12. You can probably do this in your head for these numbers:

24" = 2 ft.

18" = 1.5 ft.

2' × 1.5' = 3 sq. ft.

Either method produces the same answer, and both are easy to do on the calculator.

Change Square Feet to Square Inches

You may need to change square feet to square inches, especially in order to find the dimension in inches for a duct if the area is given in square feet. To change square feet to square inches, multiply by 144:

Area in sq. in. = Area in sq. ft. × 144

For example, find the area in inches of a duct with an area of 3.8 sq. ft.:

Area in sq. in. = Area in sq. ft. × 144

Area in sq. in. = 3.8 sq. ft. × 144

Area = 547.2 sq. in.

PROBLEMS

3. A duct measures 30" × 16". What is its area in square feet (to two decimal places)?

4. What is the area in square feet (to two decimal places) of a duct that measures 42" x 34"?

5. What is the area in square inches of a duct that has an area of 1.75 sq. ft.?

6. If a duct has an area of 5 sq. ft., what is the area in square inches?

Check your answers at the end of the chapter.

FIND A DUCT SIDE DIMENSION

Sometimes you know what the area of a duct should be and must find the dimensions of a duct that would have that same area. For example, suppose you are installing a 24" x 24" duct. The duct dimensions must change to 14" high in order to fit between a beam and a pipe. The problem is to determine the width that the new duct should be so that the area is the same as a 24" x 24" duct.

Change the equation to solve for width:

Area = Width x Height

$$\text{Width} = \frac{\text{Area}}{\text{Height}}$$

Duct sizes are usually **even numbers**. Always round to the next larger even number, not a smaller one, because it is usually acceptable to have a slightly larger area than necessary, but not a smaller one.

EXAMPLE

A duct 32" x 32" must be changed to a height of 28" so that it can fit between a beam and a pipe. How wide should the duct be in order to keep the same area?

Step 1: Find the area.

Area = Width x Height

Area = 32" x 32"

Area = 1024 sq. in.

Step 2: Find the width of the new duct.

$$\text{Width} = \frac{\text{Area}}{\text{Height}}$$

$$\text{Width} = \frac{1024 \text{ sq. in.}}{28"}$$

Width = 36.57" (Round up to 38")

Since duct sizes are even numbers, the answer of 36.57 is rounded up to the next highest even number, which is 38".

PROBLEMS

7. A duct run is 20" x 16". A transition must be made so that the duct keeps the same area but is only 10" high. (Remember that height is the second dimension.) What should the new duct size be?

8. A duct is 20" x 8". It must keep the same area, but only be 18" wide. (Remember that width is the first dimension.) What is the new duct size?

Check your answers at the end of the chapter.

FIND THE AREA OF ROUND DUCT

The size of round duct is given by the **diameter** (Fig. 4) in inches. The **radius** is **half the diameter** (Fig. 4). Divide the diameter by 2 to find the radius. You can probably do this in your head. For example, 12" round duct has a radius of 6".

The equation to find the **area of a circle** uses the symbol π, called **pi**, and the radius:

Area of circle = $\pi \times$ Radius^2

The symbol π represents a number a little over 3. Many calculators have a key for π carried to many decimal places. If you do not have a π key on your calculator, use the number 3.1416.

The term **radius^2** means **radius squared.** To square a number is to multiply it by itself. For example, if the radius is 6", multiply that number by itself:

Radius^2 = 6 x 6

Radius^2 = 36

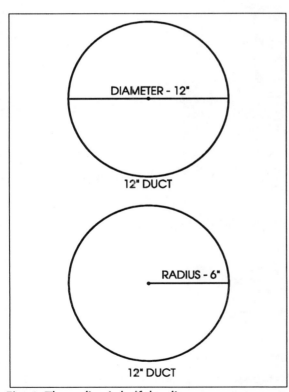

Fig. 4: The radius is half the diameter

EXAMPLE

A round duct has a 16" diameter. What is the area? (Remember that the radius is half the diameter, so the radius is 8".) If the radius is given in inches, the area is given in **square inches.**

$$\text{Area} = \pi \times \text{Radius}^2$$

$$\text{Area} = 3.1416 \times 8" \times 8"$$

$$\text{Area} = 201.06 \text{ sq. in.}$$

Just as with rectangular duct, you will often have to change the square inches to **square feet.** Divide the area by 144 to change square inches to square feet:

$$\text{Area (sq. ft.)} = \frac{\pi \times \text{Radius}^2}{144}$$

$$\text{Area (sq. ft.)} = \frac{3.1416 \times 8" \times 8"}{144}$$

$$\text{Area} = 1.4 \text{ sq. ft. (Rounded off)}$$

PROBLEMS

9. How many square **inches** (to two decimal places) are there in a 9" diameter circle?

10. How many square **feet** are there in the 9" diameter circle (to two decimal places)?

11. How many square feet (to two decimal places) are there in a 22" diameter duct?

Check your answers at the end of the chapter.

FIND THE DIAMETER OF ROUND DUCT

You may know the area of a rectangular duct in square feet and need to determine what diameter of round duct will have approximately the same area.

Change the equation for the area of a circle to solve for the radius:

$$\text{Area} = \pi \times \text{Radius}^2$$

$$\text{Radius}^2 = \frac{\text{Area}}{\pi}$$

$$\text{Radius} = \sqrt{\frac{\text{Area}}{\pi}}$$

The symbol $\sqrt{}$ means the square root of a number. The square root is the opposite of a squared number. For example, the square root of 16 is 4 because 4 squared is 16 (4 x 4 = 16). The square root of 16 is written $\sqrt{16}$:

$$\sqrt{16} = 4$$

Use a calculator to find square roots.

EXAMPLE

A rectangular duct has an area of 192 sq. in. What diameter round duct has the equivalent area?

$$\text{Radius} = \sqrt{\frac{\text{Area}}{\pi}}$$

$$\text{Radius} = \sqrt{\frac{192 \text{ sq. in.}}{\pi}}$$

Radius = 7.82"

Diameter = 7.82" x 2

Diameter = 15.64" (Round to 16")

The diameter is rounded off to the next highest even number (16").

NOTE: Round duct **over** 10" diameter is available in even whole numbers (12", 14", 16", etc.). Round duct **under** 10" diameter is available with a diameter of every whole number (9", 8", 7", etc.).

PROBLEMS

12. A round duct must have an area of 240 sq. in. What should the duct diameter be?

13. A round duct should have an area of 2 sq. ft. What should the diameter be?

14. A rectangular duct measures 18" x 12". It must be changed to a round duct of the same area. What should the diameter of the round duct be?

Check your answers at the end of the chapter.

REVIEW

If you can answer the following questions without referring to the text, you have learned the contents of this chapter. Try to answer every question before you check the answers in the back of the book.

For these problems, give duct diameters to the nearest whole number. Give duct dimensions to the nearest even number.

1. What is the area in square inches of 18" x 12" duct?

2. What is the area in square feet (to 2 decimal places) of 12" x 10" duct?

3. What is the area in square feet (to 2 decimal places) of 44" x 26" duct?

4. Use mental arithmetic to find the area in square feet of 48" x 36" duct.

5. A duct must have an area of 240 square inches. One side must measure 12". What should the dimension of the other side be (to the nearest even number)?

6. A 20" x 18" duct must be transitioned to a duct with the same area that is 14" high. What will the size of the duct be?

7. A 12" diameter round duct must be changed to a rectangular duct of the same area. The rectangular duct must be 8" high. What are the dimensions of the rectangular duct?

8. A 20" x 18" rectangular duct must be changed to round duct of the same area. What is the diameter of the round duct (to the nearest even number)?

9. Figure 5 shows a run of duct. Give the dimensions of the duct at points A, B, C, and D.

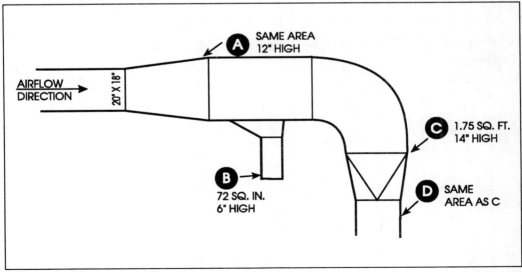

Fig. 5: Problem 9

SUMMARY OF EQUATIONS USED TO FIND DUCT SIZES

Find the area of rectangular duct in square inches:

Area = Width (in.) x Height (in.)

Find the area of rectangular duct in square feet:

$$\text{Area (sq. ft.)} = \frac{\text{Width (in.)} \times \text{Height (in.)}}{144}$$

Change square inches to square feet:

$$\text{sq. ft.} = \frac{\text{sq. in.}}{144}$$

Change square feet to square inches:

sq. in. = sq. ft. x 144

Find one side of a duct if the area and another side is known:

$$\text{Width} = \frac{\text{Area}}{\text{Height}}$$

Find the area of round duct:

$$\text{Area} = \pi \times \text{Radius}^2$$

Find the radius of round duct if the area is known:

$$\text{Radius} = \sqrt{\frac{\text{Area}}{\pi}}$$

**ANSWERS
TO PROBLEMS**
1. 720 sq. in.
2. 324 sq. in.
3. 3.33 sq. ft.
4. 9.92 sq. ft
5. 252 sq. in.
6. 720 sq. in.
7. 32" x 10"
8. 18" x 10"
9. 63.62 sq. in.
10. 0.44 sq. ft.
11. 2.64 sq. ft.
12. 18" (Remember to round up to higher even number.)
13. 20" (Remember to change the sq. ft. to sq. in.)
14. 18" (Remember to round up.)

3   AIR QUANTITY AND VELOCITY

If you work with a heating, ventilating, and air conditioning system, you need to be able to calculate the **quantity** and **velocity** of the air.

- ❑ **Air quantity** is the **volume** of air delivered by a duct in a given period of time. It is also called **rate of flow**. Air quantity is measured in **cubic feet per minute (CFM)**.
- ❑ **Air velocity** is the **speed** of air through the duct. Air velocity is measured in **feet per minute (FPM)**

All HVAC technicians must be able to calculate how much air a duct is delivering to a space and how fast the air is moving. They need to measure air delivery conditions and calculate what changes are needed.

Airflow affects comfort and indoor air quality because it delivers heated, cooled, or outside air to the conditioned spaces. Even if the air is at the proper temperature:

- ❑ If the **velocity** (speed) of the air in the duct is too fast, the system will be noisy and the conditioned space will be drafty.
- ❑ If the **quantity** of air delivered to the space is too much or too little, the space will be too hot or too cold.

When the airflow is not correct, the room is uncomfortable.

AIR QUANTITY DEPENDS ON AIR VELOCITY AND DUCT SIZE

The **quantity** of air delivered by a duct is also called the **volume**. It is measured in **cubic feet per minute**. A **cubic foot** of air would fit in a cube that is one foot on all sides (12" x 12" x 12") (Fig. 1).

It is easy to see that the quantity of air that a duct can deliver in one minute depends on two things:

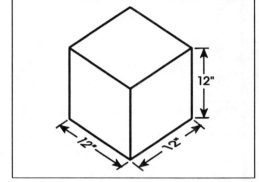

- Velocity (speed) of the air
- Size of the duct

Fig. 1: A cubic foot is one foot long on all sides

Figure 2 shows that **velocity** affects how much air is delivered:

- If duct one foot square has air flowing at **1000** FPM (feet per minute), it will deliver **1000** CFM (cubic feet of air per minute) (Fig. 2A).
- If duct one foot square has air flowing at **4000** FPM (feet per minute), it will deliver **4000** CFM (cubic feet per minute) (Fig. 2B).

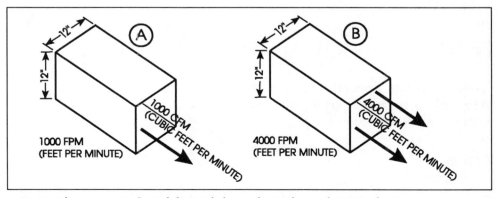

Fig. 2: The amount of air delivered depends partly on the air velocity

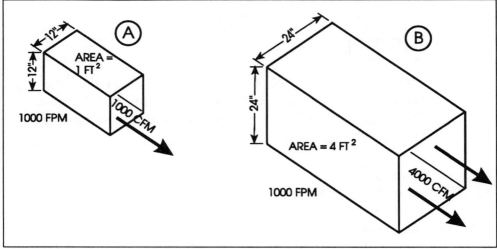

Fig. 3: The amount of air delivered depends partly on the area of the duct

Figure 3 shows that the **size** of the duct also affects how much air is delivered:

- ❑ If duct **one square foot** has air moving at 1000 FPM (feet per minute), it will deliver **1000** CFM (cubic feet of air per minute) (Fig. 3A).
- ❑ If duct **4 square feet** has air moving at 1000 FPM (feet per minute), it will deliver 4000 CFM (cubic feet of air per minute) (Fig. 3B).

FINDING AIR QUANTITY

As you can see, a small duct with the air flowing rapidly may deliver in one minute the same amount as a large duct with the air flowing more slowly. This fact gives us a very useful equation:

Quantity = Area x Velocity

EXAMPLE

Air is flowing at 1200 FPM through 28" x 18" duct. How much air is being delivered?

Step 1: Find the cross-sectional area of the duct in square feet. (This equation was covered in Chapter 2.)

$$\text{Area} = \frac{\text{Width} \times \text{Height}}{144}$$

$$\text{Area} = \frac{28" \times 18"}{144}$$

Area = 3.5 sq. ft.

Step 2: Find the air quantity.

Quantity = Area x Velocity

Quantity = 3.5 sq. ft. x 1200 FPM

Quantity = 4200 CFM

FINDING AIR VELOCITY

The equation can be written in a different form to find the **velocity** if the quantity of air and the duct area are known:

$$\text{Velocity} = \frac{\text{Quantity}}{\text{Area}}$$

EXAMPLE

A system is to deliver 2110 CFM. The duct size is 20" x 16". What air velocity is needed to handle this air quantity?

Step 1: Find the area of the duct in square feet.

$$\text{Area} = \frac{\text{Width} \times \text{Height}}{144}$$

$$\text{Area} = \frac{20'' \times 16''}{144}$$

$$\text{Area} = 2.22 \text{ sq. ft.}$$

Step 2: Find the air velocity.

$$\text{Velocity} = \frac{\text{Quantity}}{\text{Area}}$$

$$\text{Velocity} = \frac{2110 \text{ CFM}}{2.22 \text{ sq. ft.}}$$

$$\text{Velocity} = 950 \text{ FPM}$$

FINDING DUCT AREA

The equation can also be changed to find the **duct area** needed if the required air quantity and air velocity are known:

$$\text{Area} = \frac{\text{Quantity}}{\text{Velocity}}$$

EXAMPLE

A system needs to deliver 8500 CFM. The velocity in the duct is to be 1600 FPM. What should the area of the duct be?

$$\text{Area} = \frac{\text{Quantity}}{\text{Velocity}}$$

$$\text{Area} = \frac{8500 \text{ CFM}}{1600 \text{ FPM}}$$

Area = 5.31 sq. ft.

PROBLEMS

Do these problems before you go on. After you work the problems, check the answers at the end of this chapter.

Standard trade practice is to:

- ❑ Round off CFM or FPM to the nearest 5.
- ❑ Round off dimensions for rectangular duct to the nearest even number.

1. Air is moving at 1500 FPM through 32" x 18" duct. How much air is being delivered?

2. Duct that is 30" x 16" is delivering 4035 CFM. What is the air velocity?

3. If the air velocity is 1400 FPM and the duct is 24" x 24", how much air is being delivered?

4. How **wide** should a duct be if 5200 CFM must be delivered at 1500 FPM? The duct must be 18" high.

AIR VELOCITY AND AIR QUANTITY OF ROUND DUCT

The equations for air velocity and air quantity are used for round as well as rectangular duct.

As you learned in Chapter 2, the equation for finding the cross-sectional area for round duct is this:

$$\text{Area} = \pi \times \text{Radius}^2$$

Once you have calculated the cross-sectional area of round duct, you can use the three versions of the airflow equation:

$$\text{Quantity} = \text{Area} \times \text{Velocity}$$

$$\text{Velocity} = \frac{\text{Quantity}}{\text{Area}}$$

$$\text{Area} = \frac{\text{Quantity}}{\text{Velocity}}$$

EXAMPLE—FINDING AIR QUANTITY

Air is flowing at 1200 FPM through a 22" diameter duct. How much air is being delivered?

Step 1: Find the area of the round duct in square feet.

$$\text{Area} = \frac{\pi \times \text{Radius}^2}{144}$$

$$\text{Area} = \frac{3.1416 \times 11 \times 11}{144}$$

$$\text{Area} = 2.64 \text{ sq. ft.}$$

Step 2: Find the air quantity.

$$\text{Quantity} = \text{Area} \times \text{Velocity}$$

$$\text{Quantity} = 2.64 \text{ sq. ft.} \times 1200 \text{ FPM}$$

$$\text{Quantity} = 3168 \text{ CFM (Round to 3170 CFM)}$$

EXAMPLE—FINDING AIR VELOCITY

A system is to deliver 2110 CFM. The round duct has a diameter of 18". What air velocity is needed to deliver this air quantity?

Step 1: Find the area of the duct in square feet.

$$\text{Area} = \frac{\pi \times \text{Radius}^2}{144}$$

$$\text{Area} = \frac{3.1416 \times 9 \times 9}{144}$$

$$\text{Area} = 1.767 \text{ sq. ft.}$$

Step 2: Find the air velocity.

$$\text{Velocity} = \frac{\text{Quantity}}{\text{Area}}$$

$$\text{Velocity} = \frac{2110 \text{ CFM}}{1.767 \text{ sq. ft.}}$$

$$\text{Velocity} = 1194 \text{ FPM} \quad (\text{Round to } 1195)$$

EXAMPLE—FINDING THE AREA AND DIAMETER

A system needs to deliver 8500 CFM. The velocity in the duct is to be 1600 FPM. What should the diameter of the round duct be?

Step 1: Find the area in square inches.

$$\text{Area} = \frac{\text{Quantity}}{\text{Velocity}}$$

$$\text{Area} = \frac{8500 \text{ CFM}}{1600 \text{ FPM}}$$

Area = 5.31 sq. ft.

Area (in sq. in.) = 5.31 sq. ft. × 144

Area = 765 sq. in.

NOTE: You can do this in one operation on your calculator:

$$\text{Area} = 144 \times \frac{8500 \text{ CFM}}{1600 \text{ FPM}}$$

Area = 765 sq. in.

Step 2: Find what diameter has approximately that area.

$$\text{Radius} = \sqrt{\frac{\text{Area}}{\pi}}$$

$$\text{Radius} = \sqrt{\frac{765}{\pi}}$$

Radius = $\sqrt{243.5}$

Radius = 15.6

Diameter = 15.6 × 2

Diameter = 31.2" (Round up to 32")

NOTE: You can do the two parts of this process in one operation on your calculator:

$$\text{Diameter} = \sqrt{\frac{5.31 \text{ sq. in} \times 144}{\pi}} \times 2$$

Diameter = 31.2" (Round up to 32")

PROBLEMS

For these problems:

- ❑ Round off CFM or FPM to the nearest 5.
- ❑ Round up duct diameter to the next even number for duct over 10". Round up to the next whole number for duct under 10".

5. If the air velocity is 1255 FPM and the round duct has a diameter of 26", how much air is being delivered?

6. If 30" diameter duct delivers 7480 CFM, what is the air velocity?

7. What size round duct is needed to handle 1200 CFM at 800 FPM?

Check your answers at the end of the chapter.

REVIEW

If you can answer the following questions without referring to the text, you have learned the contents of this chapter. Try to answer every question before you check the answers in the back of the book.

Follow trade practice in rounding off.

1. What does CFM mean?

2. What does FPM mean?

3. A 12" x 12" duct has an air velocity of 2000 FPM. What is the CFM?

4. A 24" x 12" duct delivers 4000 CFM. What is the velocity of the air in FPM?

5. A duct measures 24" x 16". The air velocity in the duct is 1500 FPM. What is the CFM (to the nearest 5)?

6. A duct is 32" x 18". It is delivering 4800 CFM. What is the velocity in FPM (to the nearest 5)?

7. A duct must deliver 6000 CFM at 2000 FPM. The rectangular duct must be 18" high. What are the dimensions of the duct?

8. If the duct in item 7 must be round, what is the duct diameter (to the nearest whole number)?

9. The air in 22" diameter round duct has a velocity of 1550 FPM. What is the CFM?

10. A 20" round duct delivers 6000 CFM. What is the air velocity?

11. What diameter round duct is needed to deliver 2500 CFM at 1300 FPM?

12. If the duct in item 11 must be rectangular with one side 14", what is the duct size?

SUMMARY OF EQUATIONS USED WITH AIR QUANTITY AND VELOCITY

Find air quantity:

$$\text{Quantity} = \text{Area} \times \text{Velocity}$$

Find air velocity:

$$\text{Velocity} = \frac{\text{Quantity}}{\text{Area}}$$

Find duct area if air quantity and velocity are known:

$$\text{Area} = \frac{\text{Quantity}}{\text{Velocity}}$$

ANSWERS TO PROBLEMS
NOTE: Do all your calculations on the calculator and do not round off until the final answer. If you round off numbers at various points in the calculation, your answers may be a little different from the answers given below.
1. 6,000 CFM
2. 1,210 FPM
3. 5600 CFM
4. 28"
5. 4625 CFM
6. 1525 FPM
7. 18" diameter (Remember to round up to the next even number.)

4 PRESSURES IN A DUCT

You now understand how air velocity (FPM) and the duct area affect the quantity of air (CFM) that flows in a duct. You have also learned that there must be a difference in pressure (measured in "wg) for air to flow in ducts. But there are still more factors that affect airflow in a duct.

DUCT PRESSURES

Whenever air flows in a duct, three air pressures in the duct are related to the airflow:

- ❏ Static pressure
- ❏ Velocity pressure
- ❏ Total pressure

Two factors cause loss of pressure (resistance) in a duct system:

- ❏ Friction loss
- ❏ Dynamic loss

This chapter explains these three kinds of pressure when air flows in ducts. It also explains how pressure losses occur and how they affect the airflow.

Atmospheric Pressure

Consider a piece of straight duct that is open on the ends (Fig. 1). No air flows through the duct because there is no pressure difference to cause the air to flow. There is

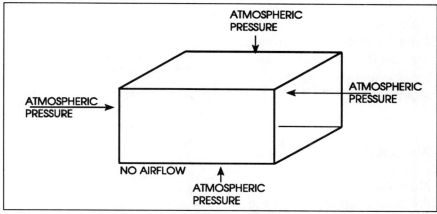

Fig. 1: Atmospheric pressure is the ordinary pressure of the air

pressure on the duct, but it is the same pressure on both ends, on the outside, and on the inside of the duct. This is the ordinary pressure of the air called **atmospheric pressure.**

Atmospheric pressure is 14.7 psi (pounds per square inch) at sea level. As you rise above sea level, there is less air above you and therefore a lower atmospheric pressure. At 5,000 feet the atmospheric pressure is normally 12.25 psi.

Atmospheric pressure is always present. It is regarded as the **zero point** for duct pressures:

- ❑ A positive duct pressure (+) is greater than atmospheric pressure. A duct pressure of 0.10" wg means that the pressure is 0.10" wg **above** atmospheric pressure.

- ❑ A negative duct pressure (-) is less than atmospheric pressure. A duct pressure of -0.10" wg means that the pressure is 0.10"wg **below** atmospheric pressure.

Since atmospheric pressure is the zero point, it does not affect the calculations in this book. However, the difference in atmospheric pressure at different elevations above sea level is a factor in more advanced calculations that will be covered in a later book in this *Indoor Environment*

Technician's Library. These calculations require knowing the weight of a cubic foot of air. Air weighs less at higher elevations.

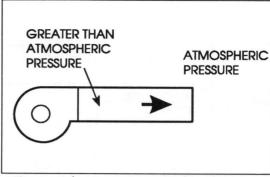

Fig. 2: Higher pressure created by the fan causes the air to flow

Velocity Pressure

Now add a fan to one end of the duct (Fig. 2). When the fan operates, the fan blades compress the air slightly by putting pressure on it. Air has weight. Just as it requires pressure to push a car, it requires pressure to push air.

Since the air at the fan outlet (the duct inlet) is under pressure which is greater than atmospheric pressure, the air moves from the higher pressure to the lower pressure at the open end of the duct (Fig. 2).

The pressure that is a result of the air in motion in the duct is called **velocity pressure.** *Velocity* means speed. The greater the velocity of the air, the higher the velocity pressure will be.

In the open air, when you feel wind in your face, it is the velocity pressure of the air that you feel. Velocity pressure created by the wind makes tree branches sway and makes a kite fly.

Static Pressure

The air pushed by the fan also exerts pressure on the sides of the duct. This is called the **static pressure.** Static means *not moving.* Compare this to a balloon (Fig. 3). A balloon stays

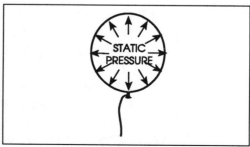

Fig. 3: Static pressure keeps a balloon inflated

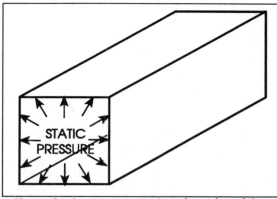

Fig. 4: Static pressure against the sides of the duct does not move the air

inflated because the air inside it presses against it on all sides. In the same way, some of the pressure from the fan is static pressure against the sides of the duct (Fig. 4).

If you release an inflated balloon, it flies through the air until the balloon is out of air (Fig. 5). This is because the static pressure in the balloon has been released. The static pressure turns to velocity pressure as it leaves the balloon. The reaction to velocity pressure propels the balloon through the air until all the static pressure is released.

Static pressure is an important consideration in the construction of a duct. There have been cases where a duct was made of light gage material that was not strong enough

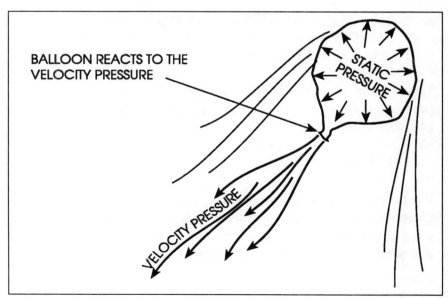

Fig. 5: Velocity pressure moves a balloon

to resist the static pressure produced by the fan. Too much static pressure ballooned the walls of the duct (Fig. 6) and caused damage.

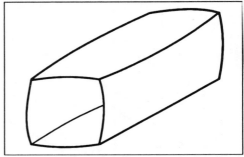

Fig. 6: Static pressure will make the sides of the duct balloon out if the duct is not strong enough

Total Pressure

Total pressure in the duct is the sum of the velocity pressure and the static pressure. This means that we deal with three kinds of air pressure in the duct:

- ❑ **Static pressure (SP)**—The pressure on the sides of the duct.
- ❑ **Velocity pressure (VP)**—The pressure that is the result of air in motion in the duct.
- ❑ **Total pressure (TP)**—The total pressure on the air. It is the sum of static pressure and velocity pressure at any one location.

These three kinds of pressure are closely related. Total pressure (TP) equals velocity pressure (VP) plus static pressure (SP):

TP = VP + SP

If total pressure remains constant, when one of the two changes, the other changes in the opposite direction. For example, with TP constant, when VP increases, SP decreases. If SP increases, VP decreases.

PRESSURE CHANGES IN A DUCT

Resistance

Think of a fan with a five foot length of airtight straight duct attached to its discharge. The air quantity in CFM (cubic feet per minute) at the fan outlet and at the end of the duct will be the same. However, if you add a hundred feet of straight duct, there is less air volume at the end of the duct even though the fan discharge pressure remains the same. If you continue to add straight duct, at some point very little air would come out of the end of the duct. This is because there is resistance to the air movement in the duct.

Resistance to air movement in a duct system is caused by both friction losses and dynamic losses due to disturbance of flow. The term **disturbance of flow** refers to any change in the direction of airflow or change in air velocity.

Friction Losses

When water flows down a stream, friction is created as the water rubs along the banks. As a result of the friction, the water along the bank flows slower than the water in the center of the stream.

In much the same way, static pressure pushes air against the sides of the duct as the air flows along. The air rubbing against the sides of the duct causes friction. This friction causes some loss of the static pressure. This is called **friction loss**.

As a result of friction loss, the total pressure is greatest at the fan outlet and gradually decreases as the air moves along the duct.

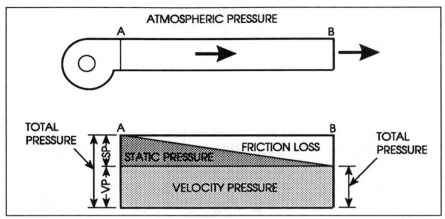

Fig. 7: Friction causes a loss of static pressure

In a straight duct with no size changes, the velocity of the air and the resulting velocity pressure remain the same. However, friction causes a loss of static pressure (SP). Figure 7 shows straight duct that is open on the end. At the fan outlet (point A) the total pressure is at its highest, and consists of both static pressure and velocity pressure. The VP remains the same throughout the duct, but friction causes a steady loss of static pressure. At the duct outlet (point B), static pressure has disappeared. At this point, total pressure is the same as the velocity pressure.

Dynamic Losses

Water flowing in a stream always has some disturbance to its flow. A large rock or tree root in a stream will make the water swirl and eddy around it. A sharp bend in the stream will cause complex currents. All these variations disturb the smooth flow of water.

Airflow in duct moves in much the same way. Dampers, coils, or other obstacles make the air swirl and eddy. An elbow breaks up the smooth flow of air. All these variations disturb the airflow (Fig. 8). Any losses due to disturbance in otherwise straight flow of air are called **dynamic losses.**

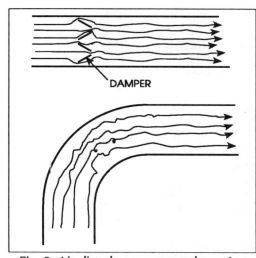

Fig. 8: Air disturbance causes dynamic loss

Dynamic losses can be caused by any change in direction or velocity in a run of duct, as a result of coils, elbows, offsets, and transitions. They all cause the air to change direction or velocity in some way. Any change in direction or velocity results in dynamic loss. Chapter 5 explains how particular fittings affect airflow.

Designers use tables to estimate how much dynamic loss a fitting will cause.

Dynamic Losses Due to Duct Size Change

Any change in duct size causes a change in velocity, so a change in size causes a dynamic loss.

Compare the airflow to a river. When a river channel narrows, the water flows faster. When the same river channel widens, the water slows down. In the same way, if duct size becomes smaller, the velocity (FPM) and the resulting velocity pressure (VP) increase. If duct size becomes larger, velocity and therefore velocity pressure decrease. Remember the equation: Quantity = Area x Velocity. If the quantity remains the same, when the area increases, the velocity must decrease. When the area decreases, the velocity must increase.

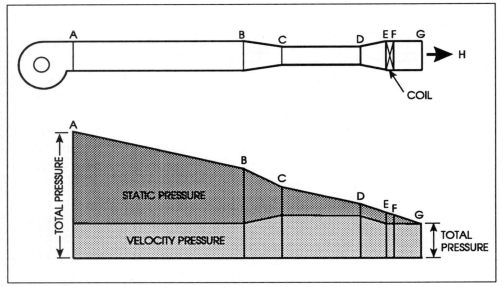

Fig. 9: Change in duct size causes change in air velocity

Figure 9 demonstrates what happens when a duct changes size:

- ❏ At point A, no friction or dynamic losses have yet occurred, so the total pressure is at its greatest. The amount of the VP depends on the size of the duct and the CFM being delivered. The rest of the total pressure is static pressure.

- ❏ From point A to point B, the duct is straight and does not change size. The VP remains the same from A to B because the air velocity does not change. The static pressure decreases from point A to point B because of friction losses. All the pressure loss is in static pressure.

- ❏ From point B to point C, the duct becomes smaller, so the VP increases (in order to deliver the same CFM.) The static pressure decreases more rapidly because of friction and dynamic losses. The result is a lower total pressure.

❏ From point C to point D the duct is straight. The VP remains the same because the air velocity remains the same. The static pressure decreases, but the loss is less than it was from B to C, where there was a size change. The total pressure decreases only a little.

PROBLEMS

The following problems are based on Fig. 9.

Problems 1 to 3 refer to the duct from points D to E:

1. What causes the loss in total pressure from point D to E?

2. Does the velocity pressure increase or decrease?

3. What causes the change in velocity pressure?

Problems 4 and 5 refer to the duct from points E to F. Note that this is a coil.

4. Why does the total pressure decrease in this section?

5. What part of the loss is in static pressure and what part is in velocity pressure?

Problems 6 and 7 refer to points G to H:

6. Why is there no static pressure at point H?

7. What happens to the velocity pressure after point G?

Check your answers at the end of the chapter.

REVIEW

If you can answer the following questions without referring to the text, you have learned the contents of this chapter. Try to answer every question before you check the answers in the back of the book.

Matching

Some items may have more than one answer.

1. Atmospheric pressure
2. Total pressure
3. Friction loss
4. Static pressure
5. Dynamic loss
6. Velocity pressure

A. Pressure on the outside walls of the duct
B. Pressure losses due to changing the direction of the air
C. The result of air in motion
D. Caused by air rubbing against the sides of the duct
E. Presses against the inside walls of the duct
F. SP + VP
G. TP - VP

7. Is the resistance to airflow greater at the fan outlet or at the end of the duct run?

8. A duct changes size from a small cross-sectional area to a much larger area. The CFM remains the same. Does the total pressure increase or decrease?

9. At a certain point in the duct run, the static pressure is measured at 1.75" wg. The total pressure is measured at 2.00" wg. What is the velocity pressure?

ANSWERS TO PROBLEMS

Your answers need not be the exact wording of those below but should contain the same general ideas.

1. Dynamic loss due to change in velocity (On this short piece, friction loss is not significant.)
2. Decrease
3. The duct size becomes larger so the air velocity decreases.
4. Because of dynamic loss due to disturbance created by the coil.
5. Most of the change is in SP.
6. As the air leaves the duct there is nothing to contain it and it spreads out into the atmosphere.
7. It gradually decreases as the air moves into the room.

5  AIRFLOW IN A DUCT AND DYNAMIC LOSSES

This chapter and Chapter 6 deal with different aspects of designing ducts and duct fittings. A **fitting** is any section of a duct run that is not straight duct. Typical fittings are elbows, offsets, and transitions.

Every technician in the HVAC industry should understand the basics covered in these chapters. This knowledge will help you deal with problems in the shop and in the field. It will also let you recognize potential problems.

The airflow in a duct is often misunderstood. If you understand the simple principles of Chapters 4, 5, and 6, you will have a much better knowledge of air handling than most technicians in the industry.

However it is important to understand that the material in these three chapters is **basic.** It will allow you to make duct fittings that keep pressure losses to a minimum; to make minor changes in a duct equipment and duct systems; and to solve problems in duct fabrication, field installation, air balancing, service work, and indoor air quality. However, complete duct system design and equipment specifications for a building are much more complex than the material presented here. Designing a ducted system should be done by a mechanical engineer who is experienced in air conditioning design.

AIRFLOW PATTERNS

The two patterns of airflow in a straight duct are laminar and turbulent:

- **Laminar flow** means air traveling in layers in a straight line (Fig. 1). Because of friction, the layer of air against the sides of the duct moves more slowly than the air in the middle of the duct.

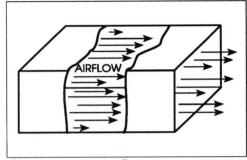

Fig. 1: Laminar airflow

- **Turbulent flow** means that the air tumbles and swirls as it moves down the duct (Fig. 2). Compare turbulent flow to water as it flows over rapids.

Laminar airflow is **stratified.** This means that the air temperatures tend to remain in layers of air and do not mix. Since the air is the medium for moving heat, this results in poor heat transportation. However, completely laminar airflow seldom occurs in duct.

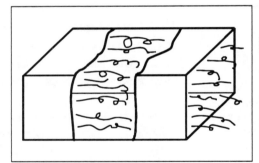

Fig. 2: Turbulent airflow

Turbulent airflow is the most common condition of airflow in duct. Turbulence helps keep the air temperature more evenly spread. However, it also results in more friction loss. Too much turbulence may create too much air noise.

The amount of turbulence in the airflow depends on a number of different factors such viscosity, friction, and air velocity. A complex calculation determines what is known as a **Reynolds number.** If the Reynolds number is above a certain value, the airflow is turbulent. For practical purposes, almost all airflow conditions in a duct are turbulent. True laminar flow occurs only in ducts that have an air velocity that is too low for practical use.

AIRFLOW AND DUCT FITTINGS

It is a law of physics that a moving object tends to travel in a straight line. It resists any change in direction. The greater the velocity, the more resistance there is to a change of direction. You have all experienced this when trying to change direction in a car that is traveling too fast. Air that is traveling also wants to travel in a straight line. The greater the velocity, the greater the resistance to changing direction.

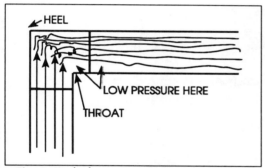

Fig. 3: Turbulence in a square throat elbow without vanes

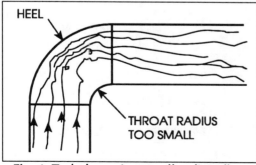

Fig. 4: Turbulence in a small radius elbow

Consider a square throat elbow without turning vanes (Fig. 3). The airflow tends to be straight until it is forced to turn. When the air hits the flat heel of the elbow, pressure build-up forces the air to turn. The result is excessive turbulence with a low pressure area at the throat of the elbow. This turbulence causes dynamic loss. A poorly designed fitting can develop a dynamic loss equal to the friction loss developed by 50 feet or more of straight duct.

A radius elbow with a throat radius that is too small (Fig. 4) has almost the same loss as a square throat elbow. The air at the heel of the elbow gradually turns, but the air near the throat tends to travel in a straight line. The result is a build-up of pressure and resulting turbulence at the heel near the end of the turn.

When a fitting creates turbulence, the turbulence continues for several feet in the straight duct that is downstream of the fitting (Figs. 3 and 4).

Duct fittings change the direction of airflow (elbows), or change the size of the duct (transitions). Duct accessories (dampers, coils) interrupt the airflow. Therefore all fittings and accessories in the duct create dynamic loss. The designer tries to keep that loss as small as practical. The general rule for fittings is to make the turns as gradual as possible and to make transitions as long as practical. "Persuade" the air to make gradual turns—you can't make it turn a square corner even if you build the duct that way.

Radius Throat Elbows

To keep air moving smoothly, a radius elbow (Fig. 5) must have a throat radius that is large enough. Design manuals recommend that a radius elbow have a **centerline radius** at least 1½ times the width of the cheek (Fig. 5). This is the same as saying that:

- ❏ **The throat radius should equal the duct width.**

For example, a duct with a cheek 12" wide should have a centerline radius of 18" (12" x 1.5 = 18"). The throat radius should be 12" (the same as the duct width).

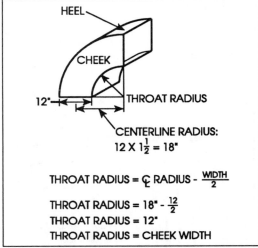

Fig. 5: Recommended throat radius

Except on a drawing, when an elbow is identified, the first dimension is the width of the cheek.

Square Throat Elbows

To be effective, square throat elbows must have turning vanes (Fig. 6). These help the air to change direction more smoothly (Fig. 7). Research has shown that **single wall vanes** (Fig. 8) produce the smallest amount of dynamic loss.

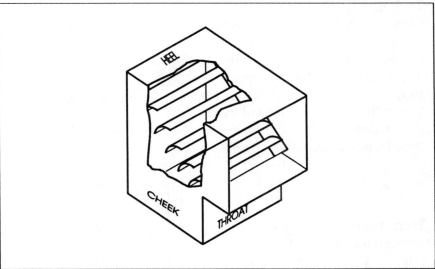

Fig. 6: Square throat elbow with turning vanes

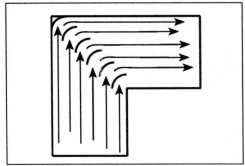

Fig. 7: Vanes allow airflow to change direction more smoothly

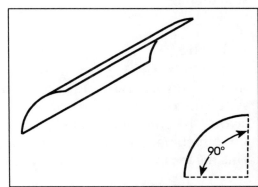

Fig. 8: Single wall turning vane

Double wall vanes (Fig. 9) were commonly used in the past. However research has shown that these are much less effective than single wall vanes.

Properly installed, a **trailing edge** on the vane (Fig. 10) will further reduce dynamic losses. However, tests have led researchers to recommend eliminating the trailing edge. This is because very often the vanes are not installed properly. The result is that they cause more loss instead of less. The vanes are effective only if they are installed carefully so that the trailing edge is parallel to the duct sides (Fig. 10). If they are not, they increase dynamic losses.

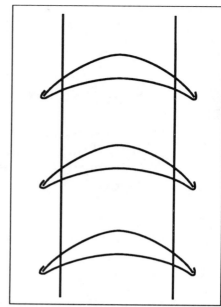

Fig. 9: Double wall vanes

Transitions

Transitions are designed to change the size of the duct (Fig. 11). To make the change gradual, the sides of transitions should be 30° or less from the straight (Fig. 11).

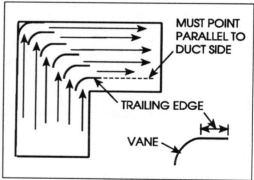

Fig. 10: Vanes with trailing edge

Offsets

Duct offsets (Fig. 12) can be made with angles or with smooth curves. The best design is the S offset (Fig. 12). When the patterns are developed by approved methods of layout, the change is gradual, the air is guided through a smooth curve, and the cross-sectional area is maintained. This means the least possible dynamic loss.

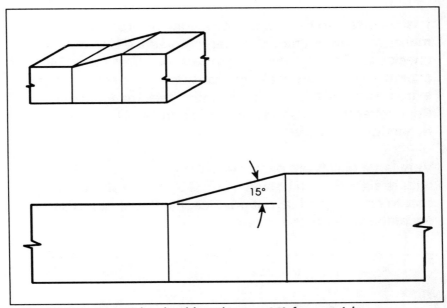

Fig. 11: Transition sides should not be over 30° from straight

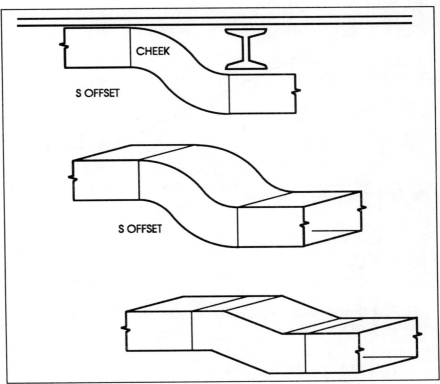

Fig. 12: Offsets

It is important to make S offsets as long as practical. This makes the curves gradual. Common practice is to cut the cheek out of a 36" sheet. This makes the offset about 34" after allowances are make for transverse connections. For extreme offsets, the cheeks are sometimes made longer. Remember that the shorter the offset, the greater the dynamic loss will be.

Very large offsets are often made using 45° or 30° elbows with straight duct in between (Fig. 13). The radius elbows should follow good layout practice: the throat radius should equal the width of the cheek.

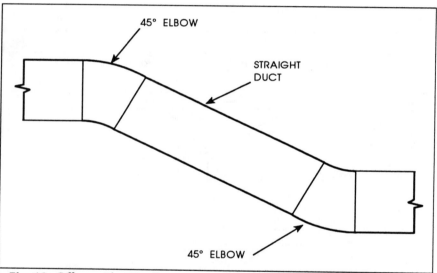

Fig. 13: Offset with 45° elbows

REVIEW

1. The throat radius of an elbow with an 18" cheek should be at least _____ inches.

Matching

2. Turbulent flow
3. Trailing edge
4. Reynolds number
5. Heel
6. Best type of turning vane
7. Transitions
8. Laminar airflow
9. Cheek
10. Dynamic loss
11. Throat

A. The flat side of an elbow
B. Air tumbles and swirls
C. The back of an elbow
D. The small side of an elbow
E. A straight length of metal on a turning vane
F. Air flows in separate layers
G. Double wall vanes
H. Single wall vanes
I. Related to the amount of turbulence in the duct
J. Changes duct size
K. Results from changing airflow direction

6 SIZING DUCTWORK

Chapter 2 showed you how to calculate duct sizes using the equation Quantity = Area x Velocity. By calculating duct sizes with this equation it is possible to change duct sizes and maintain the same quantity (CFM) and velocity (FPM) in the duct.

However, as you learned in Chapters 4 and 5, the pressure loss changes each time the duct size changes. The equation Q = A x V does not take this into account. The duct designer needs to know what the total pressure loss is for a duct run in order to select the proper size fan. The static pressure at the fan outlet must be equal to the resistance of the duct system.

Using Q = A x V to calculate duct size changes with each change in CFM maintains the same velocity, but the friction loss for each size will not remain the same. This is not to say that the equation Q = A x V is not an important equation. Understanding the relationships of this equation is essential to understanding airflow in duct.

EQUAL FRICTION LOSS METHOD OF SIZING DUCT

The industry has generally adopted the **equal friction loss method** of sizing duct. This gives the equivalent duct size based upon maintaining the same friction loss. The equal friction loss method varies the velocity but maintains the same friction loss per 100 feet of duct run. By maintaining the same friction loss per 100 feet, it is only necessary to determine the total length of the duct run to determine the total friction loss. The method is explained in this chapter.

Aspect Ratio

Understanding **aspect ratio** is important for the equal friction loss method. Choosing the best aspect ratio of a duct can reduce friction loss.

Friction loss in a duct is the result of the air molecules rubbing against the inside of the duct. For the same air quantity and velocity, a duct with a greater surface for the air to rub against will develop more friction loss. This means that if the quantity of air and the area of the duct remain the same:

- ❑ The greater the perimeter (distance around) of a duct, the more friction loss there will be.

Aspect ratio is a way to determine the best practical perimeter for a duct. Aspect ratio is the ratio between the width and height of a duct. Divide the width by the height to find the first number of the aspect ratio:

$$\text{Aspect ratio} = \frac{\text{Width}}{\text{Height}}$$

A **square duct** has an aspect ratio of 1 to 1:

$$\text{Aspect ratio} = \frac{12''}{12''} = \frac{1}{1}$$

$$\text{Aspect ratio} = 1 : 1$$

A 24" x 8" duct has an aspect ratio of 3 to 1:

$$\text{Aspect ratio} = \frac{24''}{8''} = \frac{3}{1}$$

$$\text{Aspect ratio} = 3 : 1$$

A 30" x 12" duct has an aspect ratio of 2.5 to 1:

$$\text{Aspect ratio} = \frac{30"}{12"} = \frac{2.5}{1}$$

Aspect ratio = 2.5 : 1

Figure 1 compares the aspect ratios of different size ducts that have nearly the same area (about 200 square inches) but increasing perimeters. (Actual duct is sized in even numbers. Odd numbers are used in Fig. 1 in order to maintain about the same area.) Note that the aspect ratio increases as the length of the perimeter increases. Notice also that the round pipe has the shortest perimeter and therefore the least friction loss for a given CFM and velocity.

Size	Area (sq. in.)	Perimeter	Aspect Ratio
16" Dia.	201	50"	NA
14" x 14"	196	56"	1 : 1
17" x 12"	204	58"	1.42 : 1
20" x 10"	200	60"	2 : 1
25" x 8"	200	66"	3.12 : 1
29" x 7"	203	72"	4.14 : 1
33" x 6"	198	78"	5.5 : 1

Fig. 1: Perimeters and aspect ratios of ducts with almost the same area

When using aspect ratio to choose duct sizes, follow these general principles:

- ❏ Round duct has the least friction loss.
- ❏ Next to round duct, square duct is best.

❑ As the aspect ratio increases, the friction loss increases.

❑ Avoid using duct with an aspect ratio greater than 3 to 1, if possible. In addition to increasing friction loss, it costs more to fabricate and install—more labor, more material, and heavier gages of metal.

Using the Equal Friction Loss Method

The **equal friction loss method** of sizing duct is based upon maintaining the same friction loss for every 100 feet of duct. By maintaining the same friction loss per 100 feet, it is only necessary to determine the total length of the duct run to determine the total friction loss. For example, if the duct is sized to maintain a friction loss of 0.1" wg per 100 feet and the total length of the duct run is 225 feet, calculate the friction loss:

$$0.1" \times \frac{225}{100} = 0.225" \text{ wg}$$

The duct designer can use the friction loss of 0.225" wg for the duct run to determine the size of fan needed for the system.

In order to size duct for the friction loss chosen, an equal friction chart (Fig. 2) or a duct calculator (described on page 67) is used.

To illustrate the use of the friction chart, Fig. 3 shows a run of duct supplying 8000 CFM. The CFM of each branch run is given. Assume that a friction loss of 0.1" wg per hundred feet has been selected for sizing the ductwork for the run.

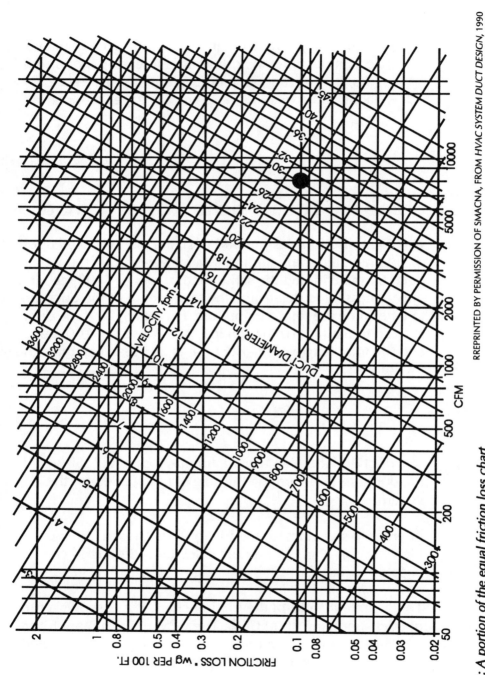

Fig. 2: A portion of the equal friction loss chart

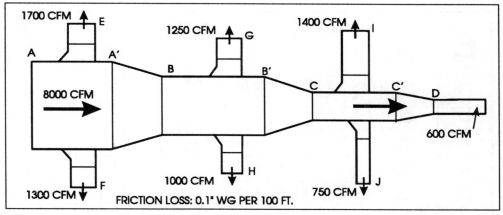

Fig. 3: Duct run

Data for Point A

The first step is to determine the duct size and the velocity at point A.

- Locate 8000 CFM on the scale at the bottom of the friction chart (Fig. 2).
- Locate 0.1" wg friction loss on the scale at the left side of the chart.
- Follow the vertical 8000 CFM line up and follow the horizontal 0.1" wg line across to where they intersect. This is marked on the chart.
- The marked point indicates that the velocity (the diagonal lines slanting down to the right) is approximately 1640 FPM.
- The marked point indicates that the duct should be the equivalent of 30" diameter (the diagonal lines slanting up to the right).

The marked point requires a 30" diameter duct, delivering 8000 CFM at 1640 FPM. If the duct run is to be rectangular, the 30" diameter must be converted to a rectangular size that will give the same friction loss. (It will not be an equal area.)

Duct Diameter in.	Rectangular size, in.	Aspect Ratio													
		1.00	1.25	1.50	1.75	2.00	2.25	2.50	2.75	3.00	3.50	4.00	5.00	6.00	7.00
9	Width	8	9	11	11	12	14								
	Height	8	7	7	9	6	6								
10	Width	9	10	12	12	14	14	15	17						
	Height	9	8	8	7	7	6	6	6						
11	Width	10	11	12	14	14	16	18	17	18	21				
	Height	10	9	8	8	7	7	7	6	6	6				
12	Width	11	13	14	14	16	16	18	19	21	21	24			
	Height	11	10	9	8	8	7	7	7	7	6	6			
13	Width	12	14	15	16	18	18	20	19	21	25	24	30		
	Height	12	11	10	9	9	8	8	7	7	7	6	6		
14	Width	13	14	17	18	18	20	20	22	24	25	28	30	36	
	Height	13	11	11	10	9	9	8	8	8	7	7	6	6	
15	Width	14	15	17	18	20	20	23	25	24	28	28	35	36	42
	Height	14	12	11	10	10	9	9	9	8	8	7	7	6	6
16	Width	15	16	18	19	20	23	23	25	27	28	32	35	42	42
	Height	15	13	12	11	10	10	9	9	9	8	8	7	7	6
17	Width	16	18	20	21	22	25	25	28	27	32	32	35	42	49
	Height	16	14	13	12	11	11	10	10	9	9	8	7	7	7
18	Width	16	19	21	23	24	25	28	28	30	32	36	40	42	49
	Height	16	15	14	13	12	11	11	10	10	9	9	8	7	7
19	Width	17	20	21	23	24	27	28	30	30	35	36	40	48	49
	Height	17	16	14	13	12	12	11	11	10	10	9	8	8	7
20	Width	18	20	23	25	26	27	30	30	33	35	40	45	48	56
	Height	18	16	15	14	13	12	12	11	11	10	10	9	8	8
21	Width	19	21	24	26	28	29	30	33	33	39	40	45	54	56
	Height	19	17	16	15	14	13	12	12	11	11	10	9	9	8
22	Width	20	23	26	26	28	32	33	36	36	39	44	50	54	56
	Height	20	18	17	15	14	14	13	13	12	11	11	10	9	8
23	Width	21	24	26	28	30	32	35	36	39	42	44	50	54	63
	Height	21	19	17	16	15	14	14	13	13	12	11	10	9	9
24	Width	22	25	27	30	32	34	35	39	39	42	48	55	60	63
	Height	22	20	18	17	16	15	14	14	13	12	12	11	10	9
25	Width	23	25	29	30	32	36	38	39	42	46	48	55	60	70
	Height	23	20	19	17	16	16	15	14	14	13	12	11	10	10
26	Width	24	26	30	32	34	36	38	41	42	46	52	55	66	70
	Height	24	21	20	18	17	16	15	15	14	13	13	11	11	10
27	Width	25	28	30	33	36	38	40	41	45	49	52	60	66	70
	Height	25	22	20	19	18	17	16	15	15	14	13	12	11	10
28	Width	26	29	32	35	36	38	43	44	45	49	56	60	66	77
	Height	26	23	21	20	18	17	17	16	15	14	14	12	11	11
29	Width	27	30	33	35	38	41	43	44	48	53	56	65	72	77
	Height	27	24	22	20	19	18	17	16	16	15	14	13	12	11
30	Width	27	31	35	37	40	43	45	47	48	53	60	65	72	77
	Height	27	25	23	21	20	19	18	17	16	15	15	13	12	11
31	Width	28	31	35	39	40	43	45	50	51	56	60	70	78	84
	Height	28	25	23	22	20	19	18	18	17	16	15	14	13	12

REPRINTED BY PERMISSION OF THE AMERICAN SOCIETY OF HEATING, REFRIGERATING AND AIR-CONDITIONING ENGINEERS, ATLANTA, GEORGIA, FROM THE 1993 ASHRAE HANDBOOK—FUNDAMENTALS.

Fig. 4: A portion of an equivalent duct sizes chart

Converting to Rectangular Duct

To convert the 30" diameter to an equivalent rectangular duct size, use a chart that gives rectangular equivalents for round duct (Fig. 4):

- ❏ The numbers on the left side of the chart are the diameters of round duct.
- ❏ The numbers in the chart in groups of two are the width and height for rectangular duct.

For example, the first group of two numbers to the right of 30" diameter are 27 and 27. This means that the equivalent of 30" diameter is a square duct which measures 27" x 27".

For practical purposes, 27" x 27" is not a good choice. First, rectangular duct is usually given in even numbers. Second, a duct 27" deep will generally not fit into the available ceiling space. Moving to the right across the chart, 40" x 20" and 48" x 16" are also given as equivalents of a 30" diameter round duct. Either size is acceptable. Assume that 40" x 20" duct is the size that best fits the space available.

The aspect ratio of a 40" x 20" duct is 2 to 1:

$$\text{Aspect ratio} = \frac{40}{20} = \frac{2}{1}$$

Aspect ratio = 2 : 1

This is an acceptable aspect ratio. Therefore, the duct chosen for point A to point A' in Fig. 3 is 40" x 20".

Data for Point B

Now determine the data for point B on the duct run in Fig. 3. The CFM required for this point of the duct will be less because two branch lines have been taken off. The CFM for these two branch lines totals 3000 CFM:

1700 CFM + 1300 CFM = 3000 CFM

Therefore the quantity needed at point B is 5000 CFM:

8000 CFM - 3000 CFM = 5000 CFM

Locate the intersection on the equal friction chart (Fig. 2) for the 5000 CFM line and the 0.1" line. This point indicates a 25" diameter duct with a velocity of approximately 1450 FPM. Since the point is in between lines, the numbers must be estimated. Estimates could vary slightly, but there will be little difference in the final selection of the rectangular duct size. Duct dimensions are rounded up to even numbers.

Figure 5 shows the data for points A and B.

Point	CFM	FPM	Dia. (Round)	Rect. Duct	Aspect Ratio
A	8000	1640	30"	40" x 20"	2:1
B	5000	1450	25"	30" x 18"	1.7:1

Fig. 5: Data for points A and B

PROBLEMS
1. Complete the table for points C to J. Round off duct diameter to the nearest whole number:

Point	CFM	FPM	Dia. (Round)	Rect. Duct	Aspect Ratio
A	8000	1640	30"	40" x 20"	2:1
B	5000	1450	25"	30" x 18"	1.7:1
C				x 18"	
D				x 10"	

Point	CFM	FPM	Dia. (Round)	Rect. Duct	Aspect Ratio
E				x 10"	
F				x 8"	
G				x 8"	
H				x 8"	
I				x 8"	
J				x 8"	

Check your answers at the end of the chapter.

Duct Calculators

Duct calculators (Fig. 6) are also available to give the same results as the equal friction loss chart (Fig. 2) and the equivalent duct sizes chart (Fig. 4). There are different forms of duct calculators available. All these calculators include the following information:

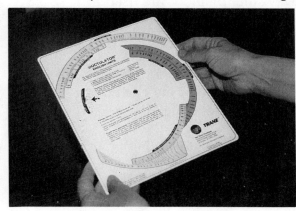

Fig. 6: Duct calculator

- ❑ CFM
- ❑ Friction loss per 100 feet of duct
- ❑ FPM
- ❑ Round duct diameter
- ❑ A selection of equivalent rectangular duct sizes

If any two of these values are known, the others can be determined.

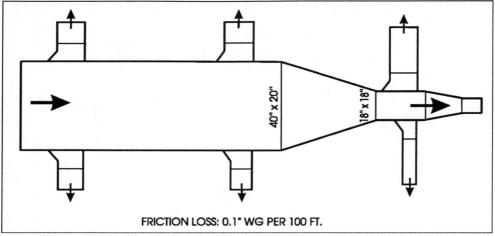

Fig. 7: Extended plenum variation of Fig. 3. It eliminates one costly transition.

OTHER METHODS OF SIZING DUCT

The equal friction loss method is the most commonly used method of sizing duct. However other methods are used.

The **extended plenum method** is a modification of the equal friction method. It eliminates some of the size changes of the equal friction method. This reduces the cost of fabrication and installation. Figure 7 shows the extended plenum design applied to the duct system in Fig. 3.

The **static regain method** reduces the air velocity in the duct. This reduces the velocity pressure. The result is that static pressure is larger than it would be under the equal friction method. (Remember that TP = SP + VP). In other words, static pressure "regains" some of its pressure.

REVIEW

1. The most common method of sizing duct is the _____ _____ loss method.

2. This method is based on the loss per _____ feet of duct run.

3. The equation for aspect ratio is _____ divided by _____.

4. A 12" x 12" duct has an aspect ratio of _____.

5. A 36" x 12" duct has an aspect ratio of _____.

6. Which has the most desirable aspect ratio, a 24" x 6" duct or a 12" x 12" duct?

Complete the chart below, based on the equal friction chart in Fig. 2.

Friction Loss	FPM	CFM	Duct Dia.
0.1" wg	800	7.	8.
9.	1200	5000	10.
11.	12.	5000	24"

13. For item 8, if one side of the rectangular duct must be 6", what is the other side?

14. For item 10, if one side of the rectangular duct must be 18", what is the other side?

15. For item 10, what size rectangular duct would have the best aspect ratio?

**ANSWERS
TO PROBLEMS**
NOTE: Your answers for FPM may vary slightly. However, this should affect the other figures little if at all.

	CFM	FPM	Dia. (Round)	Rect. Duct	Aspect Ratio
1. C	2750	1250	20"	18" x 18"	1:1
D	600	850	11"	10" x 10"	1:1
2. E	1700	1100	16"	20" x 10"	2:1
F	1300	1040	15"	24" x 8"	3:1
G	1250	1040	15"	24" x 8"	3:1
H	1000	1000	14"	20" x 8"	2.5:1
I	1400	1050	15"	24" x 8"	3:1
J	750	900	12"	14" x 8"	1.75:1

7  CALCULATING PRESSURE LOSSES IN DUCTWORK

Chapter 6 taught you how to size ducts using the equal friction loss method. This chapter explains how to calculate friction and dynamic losses, which occur in all systems.

The duct designer (usually a mechanical engineer) calculates the pressure loss in a duct system. This calculation is needed as part of the process of selecting a fan for the system. If the system pressure loss is kept low, less energy is needed to operate the fan.

It is also useful for the indoor environment technician to understand how to calculate duct pressure. This calculation is directly related to designing effective duct fittings. On installed systems, a problem can often be traced to a poorly designed duct fitting which has an excessive dynamic loss.

TOTAL PRESSURE LOSSES

The **total pressure loss** for a supply air duct system is the loss for the longest path from the outside air intake to the farthest outlet. It is **NOT** the sum of all the paths. In Fig. 1, the total system loss equals the total resistance from friction and dynamic loss of the following:

- ❑ Outside air intake
- ❑ Filters
- ❑ Coils

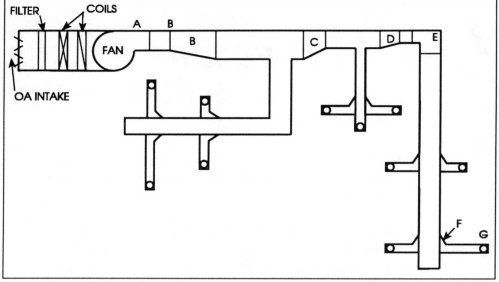

Fig. 1: System pressure loss is from OA intake to farthest outlet (G)

- Fan system effect
- Straight duct
- Transitions B, C, and D
- Elbow E
- Take-off F
- Outlet G

Fan system effect is the pressure loss created by the duct connection at the fan inlet and discharge. Fans are rated with no inlet duct and a straight outlet duct of the same dimension as the fan outlet. Any other condition affects fan performance and is part of the total pressure loss.

The **total pressure loss**, calculated from the outside air intake to the end of outlet G, is the amount of pressure that the fan must provide to overcome the system resistance.

PRESSURE LOSSES

Remember that:

- **Friction loss** is the result of air molecules rubbing on the inside surface of the duct. Friction is related to the roughness of the inside duct surface, the length of the duct, and the velocity of air in the duct. Friction loss increases greatly as air velocity increases. Sometimes it is incorrectly assumed that friction loss includes dynamic loss.

- **Dynamic loss** is the result of a disturbance to airflow in the duct. Dynamic losses occur whenever a duct changes direction or shape. The design and fabrication of a duct fitting affects the amount of dynamic loss. Changes in airflow direction or velocity should be made as smoothly and gradually as possible.

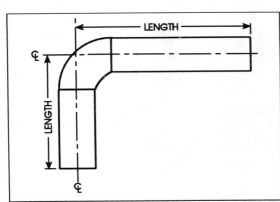

Fig. 2: Lengths of ducts are calculated from centerlines

Dynamic losses and friction losses make up the pressure losses of airflow in a duct.

LOSSES IN STRAIGHT DUCT

If the duct system has been designed by the equal friction method, calculating the friction loss for **straight duct** is simple. At elbows, the length of straight duct is measured from centerline to centerline (Fig. 2). Use this equation to calculate pressure loss for straight duct:

$$\text{Loss} = \text{"wg per 100 feet} \times \frac{\text{Length of duct (in feet)}}{100}$$

For example, suppose a straight run of duct is 322 feet long. It is designed by the equal friction method for a pressure loss of 0.15" wg per 100 feet. What is the total pressure loss?

$$\text{Loss} = \text{"wg per 100 feet} \times \frac{\text{Length of duct}}{100}$$

$$\text{Loss} = 0.15\text{" wg} \times \frac{322'}{100}$$

$$\text{Loss} = 0.15\text{" wg} \times 3.22$$

$$\text{Loss} = 0.48\text{" wg}$$

LOSSES IN FITTINGS

Fittings add additional losses. This dynamic loss is calculated separately for each fitting.

Losses for fittings used to be calculated as **equivalent duct lengths**. Tables indicated that a type of fitting had the pressure loss equal to a particular length of straight duct. For example, a table might show that the pressure loss for a certain elbow was the equivalent of the pressure loss of 70 feet of straight duct. This was added to the total length of the straight duct. If the straight duct in the system totaled 100 feet, the system loss would be calculated for 170 feet of straight duct.

A newer and better method of calculating losses for fittings is now recommended. It is based on finding the loss coefficient for each fitting from tables. A **coefficient** is simply a number determined by laboratory tests for a particular type of fitting. Once this coefficient is determined, the loss for the fitting is determined by the following equation:

$$\text{Loss} = \text{Coefficient} \times \text{Velocity pressure}$$

For example, suppose you had to find the pressure loss for the elbow in Fig. 3. The air is flowing at a velocity of 2000 FPM. Use the chart in Fig. 4, which is taken from the SMACNA manual *HVAC Systems Duct Design*, and follow these steps.

- ❑ **Identify key dimensions:** The drawing in Fig. 4 has letters to identify necessary dimensions needed to calculate the loss coefficient:

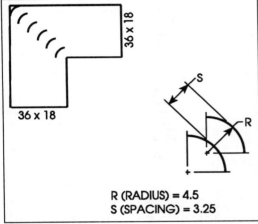

R = Radius of the vane
S = Vane spacing

Identify these dimensions on the Fig. 3 drawing:

Radius = 4.5"
Spacing = 3.25"

There are two standard sets of these radius and spacing dimensions to choose from.

Fig. 3: 36" x 18" elbow

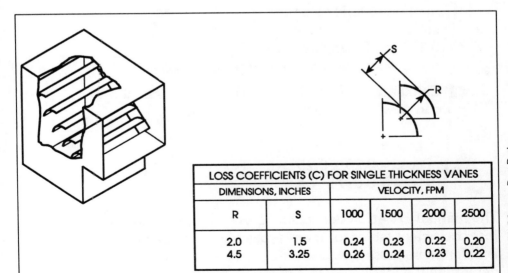

| LOSS COEFFICIENTS (C) FOR SINGLE THICKNESS VANES ||||||
| DIMENSIONS, INCHES || VELOCITY, FPM ||||
R	S	1000	1500	2000	2500
2.0	1.5	0.24	0.23	0.22	0.20
4.5	3.25	0.26	0.24	0.23	0.22

Fig. 4: Loss coefficient chart for a square elbow with vanes

- **Find the coefficient:** Next find the loss coefficient on the table in Fig. 4. Since the velocity is 2000 FPM, follow the column for 2000 FPM down to the row for 4.5 radius and 3.25 spacing. The intersection of the column and row has a coefficient of 0.23. (The coefficient is not a measurement of any sort.)

- **Find velocity pressure:** The equation for pressure loss in a fitting (Loss = Coefficient × VP) requires the figure for the velocity pressure. This can be found from a table that converts velocity (FPM) to velocity pressure (VP) (Fig. 5). The velocity in FPM for airflow in this elbow is 2000 FPM. Looking to the right of 2000 FPM on the table indicates a VP of 0.25" wg. (There are many sources for velocity/velocity pressure tables.)

Velocity FPM	Velocity Pressure in. wg.
2000	0.25
2050	0.26
2100	0.27
2150	0.29
2200	0.30
2250	0.32
2300	0.33
2350	0.34
2400	0.36
2450	0.37
2500	0.39
2550	0.41
2600	0.42
2650	0.44
2700	0.45
2750	0.47
2800	0.49
2850	0.51
2900	0.52
2950	0.54
3000	0.56

Fig. 5: A portion of a velocity and velocity pressure chart

- **Calculate the loss:** The pressure loss for the elbow can now be calculated:

    Loss = Coefficient × VP
    Loss = 0.23 × 0.25" wg
    Loss = 0.0575" wg

The manual *HVAC Systems Duct Design* by SMACNA has loss coefficient tables for many types of fittings. Each table is different because they use different ratios and methods to determine the loss coefficient of the fitting. However, each table has a drawing of the fitting to show the dimensions

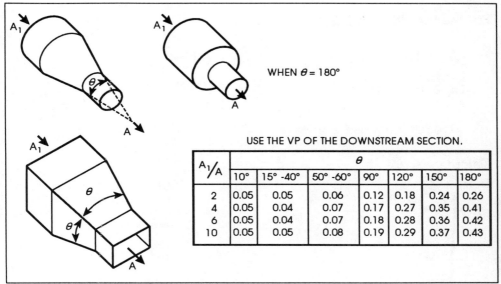

Fig. 6: Loss coefficients for a transition

needed. For example, Fig. 6 is a table for a transition with the sides slanting at the same angle:

- The letters A and A_1 stand for the area at each end of the transition.
- The symbol θ in Fig. 6 stands for the **included angle** of the tapering sides (Fig. 7). Determine the included angle by adding together the amount of taper on each side.

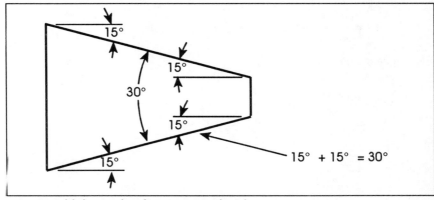

Fig. 7: Add the angle of taper on each side to determine the included angle

For the transition in Fig. 8:

$A_1 = 18" \times 18"$

$A = 9" \times 9"$

$\theta = 30°$ (because 2 sides slant 15°)

Calculate A_1/A:

$$\frac{A_1}{A} = \frac{18" \times 18"}{9" \times 9"}$$

$$\frac{A_1}{A} = 4$$

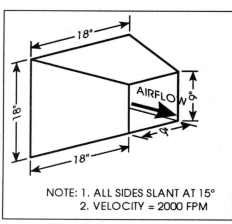

NOTE: 1. ALL SIDES SLANT AT 15°
2. VELOCITY = 2000 FPM

Fig. 8: Transition

With these figures, you can find the coefficient in the table (Fig. 6). For the included angle (30°), choose the column for 15° to 40°. Trace that column to intersect with the row under A1/A marked 4. The coefficient is 0.04.

The air velocity in the leaving duct is 2000 FPM. The table in Fig. 5 indicates that the velocity pressure for 2000 FPM is 0.25" wg.

Now you have the information needed to calculate the total pressure loss for this fitting:

Loss = Coefficient x Velocity pressure

Loss = 0.04 x 0.25" wg

Loss = 0.01" wg

SOURCES OF COEFFICIENTS

The best source of coefficient charts is *HVAC Systems Duct Design,* a manual published by SMACNA (Sheet Metal and Air Conditioning Contractors' National Association). This SMACNA manual is probably the most complete and authoritative reference on duct design. Tables for pressure loss for fittings can also be found in *Handbook of Fundamentals* published by ASHRAE (American Society of Heating, Refrigerating, and Air-Conditioning Engineers).

Pressure losses can also be calculated by computer programs. Data for each fitting and straight duct in a run is entered in the computer. The computer does all the necessary calculations and gives the total pressure loss for the complete run.

PROBLEMS

1. Calculate the pressure loss for the elbow shown in Fig. 9.

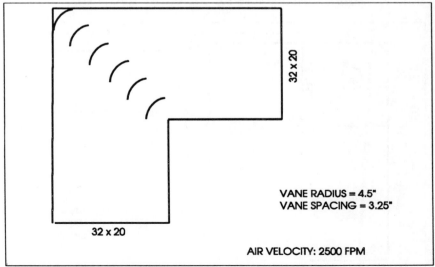

Fig. 9: Calculate the pressure loss

2. Give the data for each of the items below for Fig. 10.

   A. Total feet of straight duct for the friction loss calculation. (Do not include the transition.)

   B. **Pressure loss for straight duct**

   C. R and S for the elbow

   D. Coefficient for the elbow

   E. VP for the elbow

   F. **Pressure loss for the elbow**

   G. A_1/A for the transition

   H. θ for the transition

   I. Coefficient for the transition

   J. **Pressure loss for the transition** (For a transition, use the VP for the downstream section.)

   K. **Total pressure loss for the entire duct run**

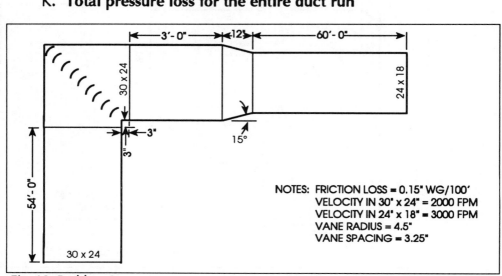

Fig. 10: Problem 2

REVIEW

1. Air rubbing against the sides of the duct creates _____ loss.

2. Anything that creates a disturbance of the airflow in a duct creates _____ loss.

3. The friction loss for a duct system is 0.15" wg per 100 ft. The length of straight duct in the system totals 160 feet. What is the total friction loss for the straight duct?

4. Calculate the friction loss for this elbow.

VANE RADIUS = 4.5"
VANE SPACING = 3.25"
24 x 18
AIR VELOCITY: 2000 FPM

5. For the duct below, determine the following:
   A. Pressure loss for the straight duct.
   B. R and S for the elbow
   C. Coefficient for the elbow
   D. VP for the elbow
   E. Pressure loss for the elbow
   F. A_1/A for the transition
   G. θ for the transition
   H. Pressure loss for the transition
   I. Total pressure loss for entire duct run

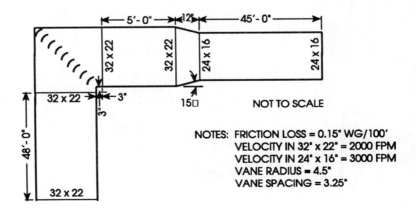

NOT TO SCALE

NOTES: FRICTION LOSS = 0.15" WG/100'
VELOCITY IN 32" x 22" = 2000 FPM
VELOCITY IN 24" x 16" = 3000 FPM
VANE RADIUS = 4.5"
VANE SPACING = 3.25"

**ANSWERS
TO PROBLEMS**
1. 0.0858" wg
   [Coefficient = 0.22
   VP = 0.39" wg
   Loss = Coefficient x VP
   Loss = 0.22 x 0.39" wg]
2A. 120
   [54' + 3" + 1' - 3" (half of 30")
   + 1' - 3" + 3" + 3' + 60']
2B. 0.1815" wg loss for straight duct

   $$[\text{Loss} = \frac{121}{100} \times 0.15" \text{ wg}]$$

2C. R = 4.5
   S = 3.25
2D. 0.23
2E. 0.25" wg
   [2000 FPM = 0.25" wg VP]
2F. 0.0575" wg loss for elbow
   [Loss = Coefficient x VP
   Loss = 0.23 x 0.25" wg]
2G. 1.667

   $$[\frac{A_1}{A} = \frac{30" \times 24"}{24" \times 18"}$$

   $$\frac{A_1}{A} = 1.667]$$

2H. 30°
   [15° + 15° = 30°]
2I. 0.05
2J. 0.028" wg
   [Loss = Coefficient x VP
   Loss = 0.05 x 0.56]
2K. 0.267" wg
   [0.1815" wg for straight duct
     0.0575" wg for elbow
    +0.028" wg for transition
     0.267" wg

8 DUCT FITTINGS

You have already learned how air flows in ducts, what pressures and pressure losses occur in duct, and how duct systems are designed. This chapter applies the knowledge to duct fittings. You will learn to reduce resistance in fittings and to recognize duct system problems.

If an HVAC system is not producing the volume of air needed, the cause can be poor fittings. You should be able to identify poor fitting applications. The chapter does not deal with the engineering aspects of duct design. It explains design in terms of practical applications for duct fittings for indoor environment technicians who service, adjust, or operate HVAC systems.

PRESSURE DROP IN FITTINGS

Any duct fitting that changes the direction of airflow or that changes the size of the duct is a potential source of large dynamic losses.

For example, in Problem 2F in Chapter 7, the pressure loss for the elbow was determined to be 0.0575" wg. The friction loss for the straight duct was 0.15" wg per 100 feet of duct. From this, it can be calculated that the friction loss of the elbow is equal to 38 feet of straight duct:

$$\frac{0.0575" \text{ wg}}{0.15" \text{ wg}} \times 100 \text{ feet} = 38 \text{ feet (38.333 rounded off)}$$

Consider the duct elbow in Fig. 1. It is a poorly designed fitting because the throat radius is too small. The duct system has a friction loss of 0.1" wg per 100 feet of straight duct. The pressure loss for this elbow with a 10" throat radius is 0.0672" wg. This equals the loss for 67 feet of straight duct.

Since any duct fitting causes dynamic losses, it is important to keep these losses to a practical minimum. Even properly designed duct fittings impose substantial pressure losses. Poorly designed fittings have much larger losses.

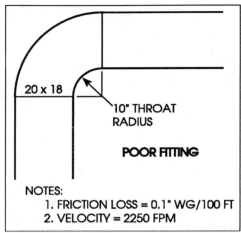

Fig. 1: Elbow with a small throat radius

On some jobs the standards for duct fittings are carefully specified. On other jobs the shop or the individual sheet metal worker makes decisions regarding throat radius or length. These decisions determine whether a fitting has a large pressure loss or a comparatively small one.

If a duct line is not delivering enough air, look at the design of the fittings to see if they might be the cause of unusual dynamic losses.

GOOD PRACTICES FOR FITTING DESIGN

If you understand the principles of good duct design, you can often locate problem areas in the duct system. There are important practices that should be observed when fabricating or installing duct.

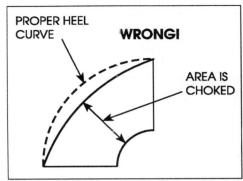

Fig. 2: Choking means reducing the fitting area

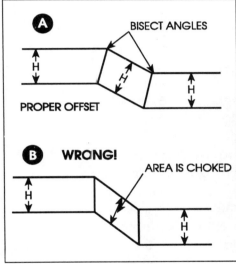

Fig. 3: A dog-leg offset must be carefully laid out to avoid choking

Avoid Choking a Fitting

A fitting is **choked** if the area in the middle of the fitting is less than the area of the ends (Fig. 2). This is the result of poor pattern drafting.

Dog-leg offsets (Fig. 3) should be avoided:

- ❏ They create a high dynamic loss.
- ❏ They can easily be choked if not laid out properly.

Dog-legs are seldom fabricated in the shop. Sometimes they are cut in the field as an emergency measure, but it is not good practice. If a dog-leg offset must be fabricated in an emergency, the angle must be bisected (Fig. 3A). If proper methods are not followed and the angle is not bisected, the area of the straight duct will be smaller than the rest of the duct (Fig. 3B).

An S offset should be used instead of a dog-leg offset.

If proper pattern drafting methods are followed for all fittings, there will be no choking. Computerized layout programs never choke a fitting.

Material	Rating	Roughness compared to aluminum
Aluminum	Smooth	x1
Galvanized steel	Medium smooth	x3
Rigid fibrous glass	Medium rough	x10
Rigid fibrous liner	Medium rough	x10
Flexible	Rough	x100
Concrete	Rough	x100

Fig. 4: Comparing the roughness of duct lining

Keep Duct Linings Smooth

Friction loss is the result of air rubbing against the duct surface. The smoother the inner surface of the duct, the less friction will be developed. If an installed HVAC system does not deliver enough air, one of the reasons could be excessive friction loss because of the roughness of the duct.

The table in Fig. 4 shows the comparative roughness of some of the commonly used duct materials. Aluminum is the smoothest of these, so it has a value of 1. All the other materials are compared to it. Galvanized steel is most commonly used for duct. For special applications, other metals such as aluminum, stainless steel, and copper are used.

Flexible duct (Fig. 5) is comparatively rough on the inside. This is one of the reasons why most

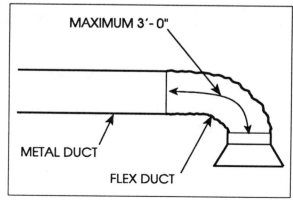

Fig. 5: The length of flexible duct that should be used is limited because it causes so much friction loss

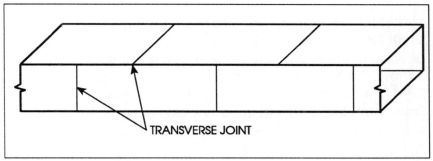

Fig. 6: Transverse joints

building codes limit the lengths of runs for flexible duct. The job specifications for the duct shown in Fig. 5 limit flexible duct to 3 foot lengths. Flex duct is commonly used to connect diffusers to ductwork to cut down labor costs.

On most jobs the material as well as the construction standards for the duct are specified. Usually SMACNA duct standards are followed.

Avoid Air Leakage

Air leakage from seams and joints in the duct system can be a major cause of energy loss. Even with careful workmanship, an unsealed system can leak as much as 30% of the total CFM.

Leakage can be controlled by careful workmanship and by sealing all **transverse joints** (Fig. 6) with a sealant. Patented duct connectors (Fig. 7) provide the tightest air seal for transverse joints. These require gaskets or caulking for the tightest seal.

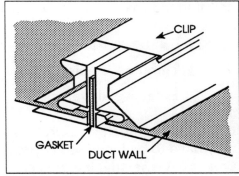

Fig. 7: Typical patented duct connector

Unfortunately, duct air leakage is commonly ignored on small jobs. On large projects, the building specifications often state the acceptable leakage as a percentage of total airflow. A commonly specified leakage rate is 5% of total CFM. Anything less than this percentage requires excessive fabrication and installation costs and is only required for special situations such as nuclear energy installations.

Avoid Heat Loss

Heat can be conducted through uninsulated metal duct walls. For heating supply duct, heat is lost to the surrounding air. For cooling supply duct, heat is gained from the surrounding air. In either case, more energy is needed to supply the conditioned space with the desired air temperature.

Supply duct in an unconditioned space, such as an attic space, should be insulated to prevent this unwanted transfer of heat. Wrap the outside of the duct with insulation to prevent heat loss or heat gain. Some duct materials, such as rigid fibrous ductboard, are insulating materials themselves and do not need additional insulation.

Avoid Air Noise

Air noise from the HVAC system that is a problem in the conditioned space can come from three sources:

- ❑ Fan noise
- ❑ Turbulence noise
- ❑ Excessive air velocity

If there is too much pressure loss in a system, the fan speed may have to be increased to provide the required static pressure. Generally, the greater the fan speed (RPM), the greater the fan noise.

Turbulence noise is produced by the fan, by improperly reinforced ductwork, and by all the duct fittings. Diffusers, registers, and grilles can also add to the noise.

Air noise can be prevented by good duct design:

- ❏ Keep the air velocity as low as practical. In general, branch lines to the outlets are designed for a lower velocity than the main duct.
- ❏ Design the duct fittings for low dynamic loss.
- ❏ Locate diffusers, registers, and grilles at least 3 feet away from any duct fittings.
- ❏ Apply acoustic lining inside the duct walls. Acoustic lining must be installed very carefully to prevent excessive friction and erosion of the material.

A developing technology is electronic noise control. The basic idea is that a microphone picks up the duct noise and the noise is analyzed by a computer. The computer generates a counter sound that is out of phase with the air noise. This sound is transmitted to a speaker in the duct. It effectively silences the air noise.

Change Duct Direction Smoothly

Like a high speed car taking a curve, air tends to move in a straight line. It resists changing direction. The result is that elbows create extra friction and dynamic loss from turbulence. Therefore whenever an elbow or offset is used to change direction, the fitting should be designed to make the airflow change direction as smoothly as possible. An offset should always be as long as practical to keep the airflow smooth.

Elbows can turn any angle, they can offset up or down, and they can change size from one end to the other.

To keep air moving smoothly and to reduce the dynamic losses, vanes are added.

Turning vanes (Fig. 8) are used in square throat elbows. As you learned in Chapter 5, laboratory tests have shown that single thickness vanes create **less** pressure loss than double thickness vanes. However, vanes longer than 36" should be reinforced or should be double thickness in order to hold their shape.

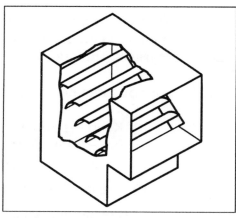

Fig. 8: Turning vanes

Turning vanes are held in **rails** (Fig. 9). Vanes and rails are purchased from manufacturers. Rails come in strips that are cut to fit the elbow. They must extend the full distance between the throat and the heel. Do not install short rails.

Pre-punched slots on the rails fit the curve of the vane. The vanes are slipped into the slots in the rails and the metal is crimped over to hold them in place. The standard spacing on these rails is $3\frac{1}{4}$" between vanes.

Vanes must always be placed in each slot. Using only every other pair of slots is a poor practice which increases dynamic losses. Tests have shown that this practice more than doubles the pressure loss for the fitting.

After the vanes are fastened in the rails, the

Fig. 9: Rail for turning vanes

rails and vanes are installed inside the elbow. The rails must be positioned so that the vanes are in line with the airflow:

- ❏ The **leading edge** must be aligned with the entering air (Fig. 10).
- ❏ The **leaving** edge must direct the air in line with the duct turn (Fig. 10).

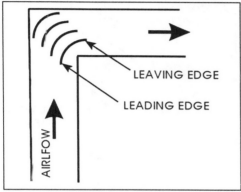

Fig. 10: Vane is aligned with airflow

If turning vanes are not positioned properly, they can create more disturbance than an elbow without vanes (Fig. 11).

Splitter vanes (Fig. 12) are used in radius throat elbows to reduce dynamic loss. A splitter vane runs the full length of the elbow. The number of splitter vanes used depends on the throat radius of the elbow. Splitter vanes for radius throat elbows are not used as often as turning vanes in square throat elbows. However, they may be required, especially if the throat radius is small.

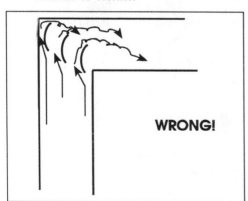

Fig. 11: Vanes out of alignment cause turbulence

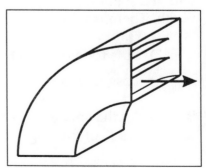

Fig. 12: Splitter vane

The throat radius is very important in determining pressure loss. For example, Fig. 13 shows two versions of the same elbow. One has a 6" throat radius and the other has a 24" throat radius:

- Loss for 6" radius elbow is 0.13" wg.
- Loss for 24" radius elbow is 0.05" wg.

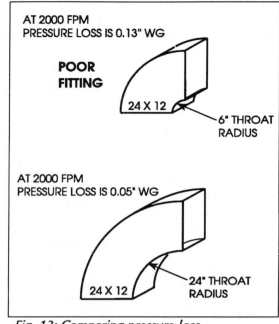

Fig. 13: Comparing pressure loss

The elbow with the 6" throat has nearly three times the pressure loss as the elbow with the 24" throat.

Splitter vanes are effective because increasing the throat radius of a radius elbow reduces the dynamic loss. The elbow in Fig. 13 with a 6" throat radius generates a large pressure loss (0.13" wg) because the throat radius is only ¼ of the cheek width. By adding two splitter vanes (Fig. 14), the elbow is, in effect, turned into three elbows, each one with a throat radius close to or greater than the cheek width. There is a large change in pressure loss:

- Loss for elbow without splitter vanes is 0.13" wg.
- Loss for elbow with 2 splitter vanes is 0.0125" wg.

Because of the splitter vanes, the pressure loss has dropped to about 10% of the original 0.13" wg.

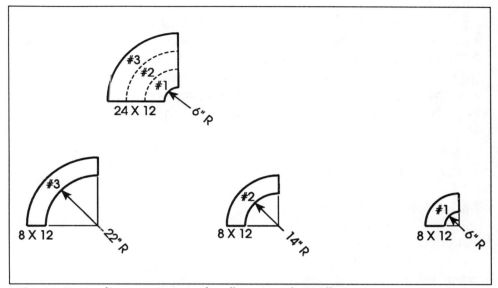

Fig. 14: Two splitter vanes turns the elbow into three elbows

Keep Transitions Gradual

To reduce dynamic loss, make transitions as long as practical. Doing this keeps the **included angle** (Fig. 15) between the two sides of a transition as small as possible. A small included angle means a small **angle of change** (Fig. 15). The included angle should not be more than 30° if possible.

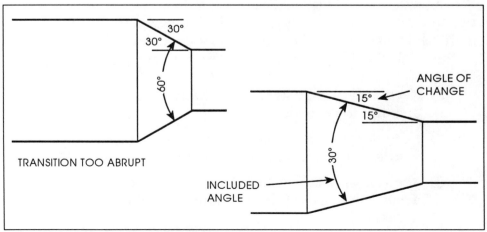

Fig. 15: Included angle and angle of change

Figure 16 shows two versions of a transition. One is short and has a 60° included angle. The other is longer and has a 30° included angle. The fitting with the small included angle (30°) has about 45% **less** pressure loss than the fitting with the large included angle (60°). Of course this assumes that both are flat on the sides and both have the same air velocity.

Take-offs

Take-offs require the air to change direction. Figure 16 shows the best design for branch take-offs that are 90° to the main duct. The length of the take-off should be ¼ of the width of the branch duct and no less than 4". It should have a 45° angle.

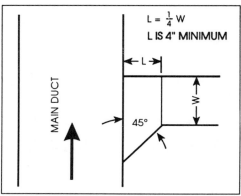

Fig. 16: Branch take-off

SUMMARY OF DUCT FITTING APPLICATIONS

Poorly designed and installed duct and fittings can ruin the performance of an HVAC system. At high pressures and high velocities, duct design is even more critical. Consider all the principles of good duct and fitting applications.

- ❏ Keep duct as large as practical for lower air velocity and thus lower losses.
- ❏ Keep duct runs as straight as possible. Each change in direction in a duct run creates a dynamic loss.
- ❏ Keep air velocity low to reduce air noise. Use acoustic lining or electronic noise control if necessary.

- ☐ Use the following equation to determine pressure loss in a fitting:

    Loss = Coefficient x VP

- ☐ Use gradual turns and changes. Make transitions and offsets as long as practical.
- ☐ Use 45° angle on take-offs.
- ☐ Use turning vanes or splitter vanes in elbows.
- ☐ Follow good pattern development practices to avoid choking fittings.
- ☐ Keep flexible duct as short as possible to avoid friction loss.
- ☐ Insulate supply duct to reduce heat loss or heat gain and save energy.
- ☐ Use careful workmanship on duct and use sealant to avoid air leakage.

REVIEW

1. If the area of a fitting is less in the middle than on the ends, the fitting is _____.

2. Flexible duct and other rough duct linings increase _____ loss.

3. A commonly specified leakage rate for a job is not to exceed _____% of total CFM.

4. List three common causes of air noise in a duct system.

5. List four ways to reduce turbulence noise.

6. The throat of a take-off should be at a _____° angle.

7. The heel of the take-off should be _____ of the width of the branch duct, and a minimum of _____ inches.

8. For the same CFM, does larger duct increase or reduce air velocity?

9. Does higher air velocity increase or reduce air noise?

10. What is the equation for pressure loss in a fitting?

9 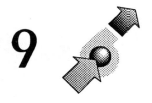 MEASURING AIRFLOW

The chapters in this book have dealt with static pressure (SP), velocity pressure (VP), air velocity (FPM), and air quantity (CFM). This last chapter explains how instruments are used to measure these quantities **in duct** and **at outlets**.

This chapter does not cover detailed instructions for measuring airflow. These will be covered in another book in this series. Measuring airflow precisely requires more knowledge and practice than one chapter can cover.

Determining the CFM in ducts, branch ducts, and outlets is one of the basic processes of the TAB (testing, adjusting, and balancing) technician. It is also essential for the IAQ (indoor air quality) technician because IAQ problems are often related to the air circulation in a room as provided by the HVAC system. The HVAC installer must know how to take airflow measurements in order to discuss problems with engineers and technicians.

The airflow quantity (CFM) in a duct is not measured directly. It is determined by going through other measurements and calculations:

- ❏ Measure VP (velocity pressure).
- ❏ Convert VP to velocity (FPM) and average the readings.
- ❏ Calculate CFM using the equation CFM = Area x Velocity.

This sounds like an involved process. However, most digital air pressure measuring instruments at least convert the velocity pressure (VP) to velocity (V). Some electronic instruments allow you to input the duct size and automatically compute the CFM.

DETERMINING AIRFLOW IN DUCT

Instruments that measure airflow in duct consist of two basic components:

- ❏ A sensing device (such as a pitot tube)
- ❏ A readout device (such as a manometer)

The **sensing device** senses the airflow pressures (velocity pressure or static pressure). The **readout device** receives the signal from the sensing device and turns it into an analog or digital reading. (An **analog** reading is a scale and indicator, such as a needle and dial. A **digital** reading is in numbers.)

Pitot Tube

The **pitot** (pea'-toe) **tube** (Fig. 1) is the most commonly used sensing device for measuring velocity pressure. It is actually two tubes, one inside the other (Fig. 2). The tube is inserted in the duct with its tip pointed into the air stream:

- ❏ **Total pressure** is sensed through the hole in the tip of the tube.
- ❏ **Static pressure** is sensed through the holes around the outside of the tube.

The pitot tube has two outlets, called **ports** (Fig. 2). One transmits total pressure (TP) and the other transmits static pressure (SP). If the **static pressure** is subtracted from the **total pressure**, the remaining

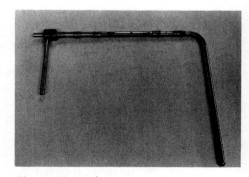

Fig. 1: Pitot tube

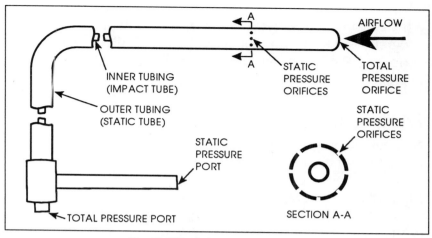

Fig. 2: Pitot tube

Fig. 3: Inclined manometer

Fig. 4: Electronic manometer

pressure is **velocity pressure** (VP). (This is based on the equation TP = VP + SP, which can be rewritten VP = TP - SP.)

Manometers

A **manometer** is usually used as the readout device that is connected to the two ports on the pitot tube—the total pressure port and the static pressure port. The manometer subtracts SP from TP to give a reading in VP.

The **inclined manometer** (Fig. 3) is generally used for very accurate readings and for calibrating other devices.

The **electronic manometer** (Fig. 4) provides a digital readout of velocity pressure or static pressure. There are different makes and models, but they commonly use a pitot tube to sense airflow pressures.

An electronic manometer is a small computer and can be programmed to do various operations. Usually it converts a series of VP readings to velocities (FPM), and provides an average of these velocities.

Magnehelic Gages

Magnehelic gages (Fig. 5) can also be used to read velocity pressures. They use special tips to sense static pressure.

Magnehelic is a trade name, but it is so widely used that it has become a generic term.

Magnehelic gages are often used to indicate differential pressure. They have sensors permanently mounted at selected places in an air handling system to show the difference in static pressure across a component. For example, the static pressure sensors of a Magnehelic gage can be mounted on each side of a bank of filters. The gage indicates the difference in pressure between the two locations. Such a differential pressure reading shows when the filters are too dirty.

Fig. 5: A Magnehelic gage and sensor

Magnehelic gages are generally not used for determining velocity pressure in a duct having a velocity less than 4000 FPM.

Taking a Pitot Tube Traverse

Because the airflow in a duct is turbulent and the velocity is not uniform, a single reading of velocity pressure at any one point is not a good indication of the velocity in the duct. Even a few random samples will not give an accurate picture of the velocity in the duct. Therefore several readings are taken in a single cross section and the readings are converted to velocity in FPM and averaged. The location and the process for the readings is called a **pitot tube traverse.**

There must be a systematic method of taking these readings. The duct cross section chosen should be at a spot where the airflow velocity is as uniform as possible. Therefore it should be as far downstream of fans and fittings as is practical. This should be at least 7.5 duct diameters downstream and 2.5 diameters upstream of any disturbance caused by a fitting.

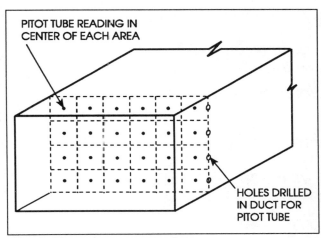

Fig. 6: Pitot tube traverse points

Square Duct Traverse

For a **square duct traverse** (used for any rectangular duct), the cross section of the duct is divided into an imaginary grid with squares of approximately 6" x 6" (Fig. 6). The number of squares depends upon the size of the duct. The general rule is that there should be no fewer than 16 squares and no more than 64. Details of the layout of the traverse squares is covered in another book in this series.

A pitot tube reading is taken in the center of each square. Holes are drilled in the duct at the proper locations to insert the pitot tube (Fig. 6). These holes may be in the side or the bottom of the duct. The pitot tube is marked (Fig. 7) to be sure it is inserted the proper amount to hit the center of each square. Pitot tubes can be obtained in different lengths to reach into any size duct.

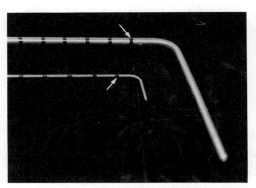

Fig. 7: Pitot tube marked for depth of insertion

Each of these velocity pressure (VP) readings is converted to velocity (FPM). Then all of the FPM readings are averaged to obtain the FPM of the airflow at that location of the duct. (The VP readings cannot be averaged because the VP is a squared function of velocity and therefore cannot be averaged.)

If only a VP reading is available, velocity (FPM) can still be determined by one of the following two methods:

- ❏ Use a table to convert VP to velocity (FPM). (There is a partial table on page 76 of this book.)
- ❏ Use the following equation to convert VP to velocity:

$$\text{Velocity} = 4005 \times \sqrt{\text{Velocity Pressure}}$$

For example, the VP is 0.25" wg. What is the velocity in FPM?

$$\text{Velocity} = 4005 \times \sqrt{\text{VP}}$$
$$\text{Velocity} = 4005 \times \sqrt{0.25\text{" wg}}$$
$$\text{Velocity} = 4005 \times 0.5$$
$$\text{Velocity} = 2002.5 \text{ FPM}$$

When the velocity is determined, the airflow quantity (CFM) is calculated with the equation CFM = Area × Velocity. If the readout device does not do this automatically, you can use your own calculator to determine CFM.

Round Duct Traverse

For a 20-point **round duct traverse**, the duct cross section is divided into five concentric circles with the pattern of points shown in Fig. 8. These two sets of traverse points must be 90° to each other. Figure 9 shows how each of these points is numbered. Points 1 to 10 are taken by

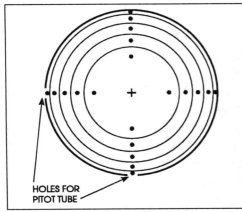

Fig. 8: Round duct traverse

inserting the pitot tube in one hole in the duct (Fig. 8). Points 11 to 20 are taken from another hole in the duct.

The traverse points are located by multiplying the **duct diameter** by the **constants** in Fig. 9. This gives the distance measured in from the wall of the duct for the pitot tube reading. For example, for a 10" diameter duct, readings 1 and 11 are taken ¼" from the hole in the duct:

- 10" × 0.026 = 0.26", which is close to ¼" from the hole in the duct.

Readings 10 and 20 are taken 9¾" from the hole:

- 10" × 0.974 = 9.74", which is close to 9¾" from the hole in the duct.

The rest of the procedure is the same as for square duct.

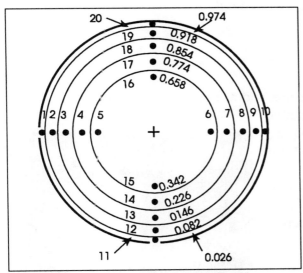

Fig. 9: Numbered points and multipliers for a round duct traverse

DETERMINING AIRFLOW AT OUTLETS

A pitot tube cannot be used to determine CFM at **outlets** (registers, grilles, and diffusers). This is because there is no ductwork, so there is no static pressure.

Direct Reading Instruments

Direct reading instruments used to be commonly used to measure the velocity at outlets. However, it is difficult to measure accurately with these instruments. The airflow is turbulent, has different spreading patterns, and is partially blocked by bars or diffuser cones. Both **rotating vane anemometers** and **velometers** provide very inaccurate readings, since the airflow is so turbulent and irregular. If dampers behind the face of the outlet are partially closed, the readings are even more inaccurate. Velometers cannot be used with registers and grilles.

In addition, a great deal of mathematics is required to determine CFM using these instruments. This takes time and the results are subject to errors in calculation.

Flow Hoods

Flow hoods are an accurate method of measuring airflow. A **flow hood** (Fig. 10) is also called a **capture hood**. It consists of a **hood**, also called a **skirt**, that fits tightly over an outlet or inlet and a meter that indicates CFM. The hood is placed on a flat surface around the outlet. A short period of time is required to obtain an accurate reading.

When the flow hood is placed over a diffuser, it creates **back pressure**, an additional pressure on the airflow from the outlet. Back pressure can introduce a large error in the CFM reading. Newer digital hoods can compensate for back pressure. On older hoods and analog

Fig. 10: Flow hood

hoods the back pressure compensation must be estimated according to the manufacturer's information.

Some electronic meters not only measure CFM, but also correct for altitude, indicate supply air temperature, provide back pressure compensation for high airflow rates, and record several readings for future recall.

Flow hoods can have analog or digital readings. A flow hood with an analog readout is faster. However, it must be corrected for air density other than standard air, and it cannot provide back pressure compensation.

A flow hood with a digital readout is a little slower, but it has excellent accuracy and it can have back pressure compensation.

Flow meters are expensive and should be treated as the fine precision instruments that they are.

REVIEW

1. The orifice at the end of the pitot tube senses _____ pressure.

2. The orifice at the side of the pitot tube senses _____ pressure.

3. For very accurate readings the _____ manometer is used.

4. A _____ gage is often used to indicate pressure on two sides of an HVAC component.

5. To take a pitot tube traverse on a rectangular duct, the cross section of the duct is divided into a grid of approximately _____" x _____" squares.

6. A pitot tube traverse usually should not have less than _____ nor more than _____ points.

7. For a 15" diameter duct, give the distance the pitot tube must be inserted in the duct for each of the following grid numbers. (Refer to Fig. 8. Answers to nearest ⅛".)
    #1. _____ "
    #7. _____ "
    #10. _____ "
    #11. _____ "
    #15. _____ "
    #20. _____ "

8. The best instrument for measuring airflow at duct outlets is a _____ _____.

9. A pitot tube traverse on a 36" x 18" duct results in the following VP readings. Calculate the velocity for each of these VP readings using the equation $V = 4005 \times \sqrt{VP}$. Round answers to the nearest whole number.
    #1. 0.32" wg
    #2. 0.33" wg
    #3. 0.32" wg
    #4. 0.34" wg
    #5. 0.32" wg
    #6. 0.29" wg
    #7. 0.32" wg
    #8. 0.34" wg
    #9. 0.42" wg
    #10. 0.47" wg
    #11. 0.36" wg
    #12. 0.29" wg
    #13. 0.30" wg
    #14. 0.30" wg
    #15. 0.29" wg
    #16. 0.32" wg
    #17. 0.29" wg
    #18. 0.27" wg

10. What is the velocity in FPM for this traverse (to the nearest whole number)?

11. Based on the velocity in item 10, what is the CFM of the duct at this traverse point? (Round off to the nearest 5.)

REVIEW ANSWERS

Chapter 1

1. Heating, ventilating, and air conditioning

2. Central air handling system
   Boiler or furnace
   Refrigeration unit
   Duct system

3. Through ductwork

4. A. Return air

   B. Exhaust air

   C. Supply air

   D. Outside air

5. To heat or cool the air

6. By a heating coil

7. The chilled water coil causes the heat to transfer from the air to the coil.

8. 14.7 psi

9. Because duct pressures are very low. If measured in psi the number would be very small and inconvenient to use.

10. Water gage

11. It makes a difference of 1" between the low side and high side of the U tube. In other words, it raises the water level ½" on one side and lowers the water level ½" on the other.

Chapter 2

1. 216 sq. in.

2. 0.83 sq. ft.

3. 7.94 sq. ft.

4. 12 sq. ft.

5. 20"

6. 26" x 14"

7. 14" x 8"

8. 22" diameter

9. A. 30" x 12"

   B. 12" x 6"

   C. 18" x 14"

   D. 18" diameter

Chapter 3

1. Cubic feet per minute

2. Feet per minute

3. 2000 CFM

4. 2000 FPM

5. 4000 CFM

6. 1200 FPM

7. 24" x 18"

8. 24" diameter

9. 4090 CFM

10. 2750 FPM

11. 20" diameter

12. 20" x 14"
    (2500 ÷ 1300 x 144 = 276.923 sq. in.
    276.923 ÷ 14" = 19.78"
    If you took the answer of 20" from item 11 and changed it to a rectangular duct, the answer would be 22.44". However the 20" diameter was the result of rounding off so using that figure is not as accurate as reworking the whole problem.)

Chapter 4

1. A

2. F

3. D

4. E, G

5. B

6. C

7. At the fan outlet

8. Decrease

9. 0.25" wg

Chapter 5

1. 18

2. B

3. E

4. I

5. C

6. H

7. J

8. F

9. A

10. K

11. D

Chapter 6

1. equal friction
2. 100
3. width, height
4. 1 : 1
5. 3 : 1
6. 12" x 12"
7. 470 CFM
8. 10" dia
9. 0.06" wg
10. 28" dia
11. 0.14" wg
12. 1600 FPM
13. 14"
14. 36"
15. 26" x 26"

Chapter 7

1. friction
2. dynamic

3. 0.24"

4. 0.058"

5. A. 0.1515" wg

   B. R = 4.5"
       S = 3.5"

   C. 0.23

   D. 0.25" wg

   E. 0.0575"

   F. 1.83

   G. 30°

   H. 0.028"

   I. 0.237" wg

Chapter 8

1. choked

2. friction

3. 5%

4. Can be in any order:
   Fan noise
   Turbulence noise
   Excessive air velocity

5. Can be in any order:
   Keep air velocity low.
   Keep dynamic loss in fittings low.
   Keep outlets at least 3 feet away from duct fittings.
   Use acoustic lining.

6. 45°

7. ¼, 4"

8. Reduce

9. Increase

10. Loss = Coefficient x Velocity pressure

Chapter 9

1. total

2. static

3. inclined

4. Magnehelic

5. 6" x 6"

6. 16, 64

7. #1. ⅝"
   #7. 3⅝"
   #10. 14⅝"
   #11. ⅝"
   #15. 5⅛"
   #20. 14⅝"

8. flow hood

9. #1. 2266 FPM
    #2. 2301 FPM
    #3. 2266 FPM
    #4. 2335 FPM
    #5. 2266 FPM
    #6. 2157 FPM
    #7. 2266 FPM
    #8. 2335 FPM
    #9. 2596 FPM
    #10. 2746 FPM
    #11. 2403 FPM
    #12. 2157 FPM
    #13. 2194 FPM
    #14. 2194 FPM
    #15. 2157 FPM
    #16. 2266 FPM
    #17. 2157 FPM
    #18. 2081 FPM

10. 2286 FPM

11. 10,285 CFM

    [CFM = Area x Velocity

    $$CFM = \frac{36" \times 18"}{144} \times 2286$$

    CFM = 10,287   (Round to 10,285)]

APPENDIX

EQUATIONS FOR DUCT SIZES

Find the area of rectangular duct in square inches:

$$\text{Area} = \text{Width (in.)} \times \text{Height (in.)}$$

Find the area of rectangular duct in square feet:

$$\text{Area (sq. ft.)} = \frac{\text{Width (in.)} \times \text{Height (in.)}}{144}$$

Change square inches to square feet:

$$\text{sq. ft.} = \frac{\text{sq. in.}}{144}$$

Change square feet to square inches:

$$\text{sq. in.} = \text{sq. ft.} \times 144$$

Find one side of a duct if the area and another side is known:

$$\text{Width} = \frac{\text{Area}}{\text{Height}}$$

Find the area of round duct:

$$\text{Area} = \pi \times \text{Radius}^2$$

Find the radius of round duct if the area is known:

$$\text{Radius} = \sqrt{\frac{\text{Area}}{\pi}}$$

Fing aspect ratio:

$$\text{Aspect ratio} = \frac{\text{Width (long side)}}{\text{Height (short side)}}$$

EQUATIONS FOR AIR QUANTITY AND VELOCITY

Find air quantity:

$$\text{Quantity} = \text{Area} \times \text{Velocity}$$

Find air velocity:

$$\text{Velocity} = \frac{\text{Quantity}}{\text{Area}}$$

Find duct area if air quantity and velocity are known:

$$\text{Area} = \frac{\text{Quantity}}{\text{Velocity}}$$

Convert velocity pressure to velocity:

$$\text{Velocity} = 4005 \times \sqrt{\text{Velocity Pressure}}$$

EQUATIONS FOR PRESSURE

Calculate pressure loss for straight duct:

$$\text{Loss} = \text{"wg per 100 feet} \times \frac{\text{Length of duct (in feet)}}{100}$$

Calculate pressure in duct:

Total pressure = Static pressure + Velocity pressure

Velocity pressure = Total pressure − Static pressure

INDEX

A   Air conditioning, 1-2
Airflow patterns, 50-57
Air leakage, 87-88
Air pressure, 6
    measured in psi, 6
    measured in "wg (inches water gage), 7
Air quantity, 26-37
Air velocity, 26-37
Air volume, *See* Air quantity
Analog and digital readings, 98
Angle of change, 93
Area of duct (cross section), 12-22, 30-31
    rectangular duct, 12-17
    round duct, 19-22
ASHRAE, *Handbook of Fundamentals*, 79, 64
Aspect ratio, 59
Atmospheric pressure, 6, 38-39

B   Boiler, 5

C   Calculating duct size, 11-25
    cross-sectional area, 12-17
    duct side, 17-18
    round duct, 19-22
Central air handling system, 2-6
Chilled water coil, 5
Chiller, 5-6
Choked fitting, 85
Coefficients for fittings, 74-79, 95
Cooling system, 5-6

D   Differential pressure readings, 100
Digital and analog readings, 98
Dog-leg offset, 85
Duct calculator, 67
Duct connectors, 87

Duct design, 49
Duct fitting design, 83-96
Duct fittings, 49-57
    definition, 49
Duct fittings, dynamic losses, 74-80
Duct lining roughness, 86-87
Duct size, *See* Calculating duct size
Dynamic loss, 38, 44-47, 49-57, 73, 84

E Elbows, 51-54
Equal friction loss method, 58-67
    SMACNA chart, 62
Equations, 116-117
Equivalent duct lengths for calculating pressure loss, 74
Equivalent duct sizes chart, ASHRAE, 64
Exhaust air, 3
Extended plenum method, 68

F Fan system effect, 72
Flexible duct, 86-87
Flow hood, 104-105
Friction, 7
Friction loss, 38, 43, 58-61, 73, 83-84

H Heat loss, 88
Heating coil, 4-5
Heating system, 4-5

I IAQ (indoor air quality), 97
Included angle, fitting, 77, 93

L Laminar flow, 50

M Magnehelic gages, 100
Manometer, 8, 99-100
Math, 11
Measuring airflow, 97-107
    at outlets, 103-105
Mixed air, 4

N Noise, HVAC system, 88-89, 94

O Offsets, 54-56
Outside air, 3

P Pitot tube, 98-103
Pitot tube traverse
    rectangular duct, 101-102
    round duct, 102-103
Pressure difference, 7
Pressure in a duct, 38-48
Pressure loss, calculating, 71-82
Pressure loss computer programs, 79

R Resistance in duct, 7, 43
Return air, 3
Reynolds number, 50
Rotating vane anemometer, 104
Round duct,
    area and diameter, 19-22
    air velocity and air quantity, 31-35

S S offset, 85
Sizing ductwork, 58-70
SMACNA, *HVAC System Duct Design*, 62, 76-77, 79
Splitter vane, 91-93
Square inches and square feet, 14-17
Static pressure, 38, 40-42, 44, 46, 58, 98-99
Static regain method, 68
Supply air, 3

T TAB (testing, adjusting, and balancing), 97
Take-off, 94
Total pressure, 38, 42, 98-99
Total pressure loss, 71-72
Transitions, 54, 93-94
Turbulent airflow, 50
Turning vanes, 53-54, 90
    leading edge and leaving edge, 91
    rails, 90

U   U-tube, 8-9

V   Velocity and velocity pressure, conversion chart, 76
Velocity pressure, 38, 40-42, 44, 46
Velometer, 104
Ventilation, 1

W   Water gage, 8

LAMA Books also publishes:

Basics of Electricity
The Indoor Environment Technician's Library
6 x 9 softcover - $17.50
Available now, **Basics of Electricity** is part of a new series of books for the indoor environment technician. Topics covered include: Electrical Safety, Basic Electricity in DC Circuits, Magnetism and Electricity, Series and Parallel Circuits, AC Circuits, Reactance, Capacitors and Capacitive Reactance, Power Factor, Electrical Work, and Transformers.

The Reading Program Books A to G
38 to 72 pages
8.5 x 11 softcover
$5.25 each book
ISBN 0-88069-007-0 (set)
A series of workbooks for remedial readers at the college level. Successfully used for years in community college reading programs. Topics include word attack, vocabulary, sentence structure, punctuation, reading for detail, paragraph structure, essay structure, and critical reading.

Two different references that cover all the occupational programs in California and the western states. Each is updated every other year.

Occupational Programs in California Public Community Colleges
176 pages 8.5 x 11 softcover
$26.50
ISSN 0731-8650

Occupational Programs in the Western States
210 pages
8.5 x 11 softcover - $27.50
ISBN 0-88069-015-1

Teach! Plain Talk About Teaching
124 pages, illustrated
7 x 9 softcover - $19.95
ISBN 0-88069-014-3
A book for instructors who teach technical subjects to adults. This guide will help you lead dynamic class discussions, engross students in learning, use media aids effectively, write tests and quizzes, and manage classroom distractions.

LAMA BOOKS ORDER FORM

SHIP ORDER TO:

NAME _____ TITLE _____

FIRM _____

STREET ADDRESS _____

CITY/STATE/ZIP CODE _____

TELEPHONE (_____) _____

MAKE CHECK PAYABLE TO: **LAMA** Books
Leo A. Meyer Associates
23850 Clawiter Road
Hayward CA 94545-1723
PHONE: 510•785•1091
FAX 510• 785•1099

DESCRIPTION	Number of Copies	Amount	Total
Please enroll me as a subscriber to the **Indoor Environment Technician's Library** and send me the latest book in the series, **Math for the Technician** (at a 20% discount). (I understand that future titles will be sent periodically on approval and I may return them if they do not fit my purposes.)		$14.00 ea.	
Math for the Technician (individual copy) *Supervisor's Guide*		$17.50 5.00	
Airflow in Ducts (individual copy) *Supervisor's Guide*		$17.50 5.00	
Basics of Electricity (individual copy) *Supervisor's Guide*		$17.50 5.00	
Occupational Programs in California Public Community Colleges		$26.50	
Occupational Programs in the Western States		$27.50	
Catalog for **The Reading Program Books A-G**		No charge	
Teach! Plain Talk about Teaching		$19.95	

Subtotal _____
CA Sales tax _____
(CA residents)
TOTAL _____

PAYMENT:

❏ CHECK ENCLOSED FOR TOTAL AMOUNT DUE

❏ PLEASE BILL ME

❏ PURCHASE ORDER ATTACHED